P9-CKA-207

AVID

READER

PRESS

In the Form *of a* Question

The Joys and Rewards *of a* Curious Life

Amy Schneider

AVID READER PRESS

New York   London   Toronto   Sydney   New Delhi

AVID READER PRESS
An Imprint of Simon & Schuster, Inc.
1230 Avenue of the Americas
New York, NY 10020

First Avid Reader Press hardcover edition October 2023

AVID READER PRESS and colophon are trademarks of Simon & Schuster, Inc.

For information about special discounts for bulk purchases, please contact Simon & Schuster Special Sales at 1-866-506-1949 or business@simonandschuster.com.

The Simon & Schuster Speakers Bureau can bring authors to your live event. For more information or to book an event, contact the Simon & Schuster Speakers Bureau at 1-866-248-3049 or visit our website at www.simonspeakers.com.

Interior design by Carly Loman

Manufactured in the United States of America

10 9 8 7 6 5 4 3 2 1

Library of Congress Cataloging-in-Publication Data

Names: Schneider, Amy, 1979– author.
Title: In the form of a question : the joys and rewards of a curious life / Amy Schneider.
Description: First Avid Reader Press hardcover edition. | New York : Avid Reader Press, 2023.
Identifiers: LCCN 2023023214 (print) | LCCN 2023023215 (ebook) | ISBN 9781668013304 (hardcover) | ISBN 9781668013311 (paperback) | ISBN 9781668013328 (ebook)
Subjects: LCSH: Schneider, Amy, 1979– | Television personalities—United States—Biography. | Sexual minorities—United States—Biography. | Jeopardy (Television program) | BISAC: BIOGRAPHY & AUTOBIOGRAPHY / LGBTQ+ | SCIENCE / Essays | LCGFT: Autobiographies.
Classification: LCC PN1992.4.S283 A3 2023 (print) | LCC PN1992.4.S283 (ebook) | DDC 791.4502/8092 [B]—dc23/eng/20230717
LC record available at https://lccn.loc.gov/2023023214
LC ebook record available at https://lccn.loc.gov/2023023215

ISBN 978-1-6680-1330-4
ISBN 978-1-6680-1332-8 (ebook)

To Genevieve

"Oh, boy! Pico de gallo! They sure don't make it like this in Ohio!"

— *"Weird Al" Yankovic*

Contents

x   Contents

Author's Note

I have many young fans, but this book isn't written for a young audience. So if you're reading this, and if your parents ever tell you that you're too young to read or watch certain things, then I'd ask you not to read this one, not without asking your parents first. I'll write one for you soon.

One of the key points I want to make in this book is that I am open to being wrong, to reconsidering my beliefs. It is my hope that, in the future, I will come to disagree with, and perhaps even disavow, some of the statements I've made in these pages, because it is my hope that I will never stop learning. I look forward to your feedback. The next book will be better.

Also: this book is entirely my own story. Of course, many other people appear in this book besides myself, although in some cases I have changed names and identifying details of people whose privacy I wish to respect. But I want to be clear that the people and events that I describe in this book are described not as they happened but as I remember them. Other people would almost certainly describe them differently, perhaps dramatically so, and the fact that I'm the one that happened to get a book deal is no reason to prioritize my recollection over theirs.

So, while I have tried to be as accurate and honest as I could, I ask you to keep in mind this story is not authoritative. It's just mine.

Oh, also: The parts where I talk about drug use and other illegal activities are all completely made up. Just for the record.

In the
Form
of a
Question

CHAPTER ONE

How Did You Get So Smart?

"How did you get so smart?" It's a question I've been asked all my life, in many variations.[1] In childhood, it was often asked with a certain amount of jealousy or disdain. One reason for that disdain was the environment I was raised in. Pride is one of the worst sins in Catholicism, and the largely German Catholic community I was part of defined "pride" broadly; so broadly, in fact, that the mere fact of being talented in some field raised suspicions. Another reason was that, well, kids can be real assholes sometimes.

It's a scary time in life when you realize that grown-ups are just people, and that they will be doing less and less to help you as you grow older. The responsibility for the outcome of your life is in your hands. So, when my peers saw me succeeding more or less effortlessly at schoolwork, while their hardest work was barely getting them by, it scared them, and they expressed that fear by showing contempt for my abilities. That meant that, from my perspective, every time I did something that my peers found

---

1 "How do you *know* all that stuff?" "So are you, like, some kind of genius?" "Do you have a photographic memory?" etc.

impressive, such as winning a spelling bee,[2] getting a perfect score on my SAT, or simply using an unfamiliar word in a sentence, they would resent me, and I could tell, even if they tried to suppress it.

I've always been a pretty empathetic person, which is a double-edged sword at times like that. So, I limited myself. By blowing off homework and studying I could bring my grades down to an approachable level.[3] And whenever the question would come along, as to why or how or where I had acquired all my knowledge, it always sounded to me like a potential attack, to be deflected however I could in the moment.

But then the day came when I was on *Jeopardy!*, every weeknight, for months, succeeding spectacularly at a well-regarded intellectual pursuit on national television. Unsurprisingly, this success prompted many, many people[4] to ask me that same old question, "How do you know all of that?"

I still didn't have a satisfactory answer. Even now, I am sometimes tempted to reply, "How does anyone know anything?" But, unless the questioner is really into Kantian epistemology,[5] that response is unlikely to drive the conversation forward. And again, due to my upbringing, I obviously couldn't just let somebody compliment me without resistance. If I went around letting people praise me willy-nilly, what would be next? Having self-worth? Pursuing my dreams? Premarital sex?!?!

---

2   Okay, okay, winning five consecutive spelling bees from 1989 to 1993.

3   Also, I could blow off homework and studying, so it was a win-win from my perspective.

4   Close friends, casual friends, coworkers, neighbors, bartenders, flight attendants, Safeway cashiers . . .

5   And of course, if they *are* into Kantian epistemology, how would I ever know? #KantJoke

To avoid such dire outcomes, I had always used one of two general approaches.

One is to attribute my intelligence to factors outside of my control. With this approach, I'll observe that I was born with a brain that, for whatever reason, retains knowledge well. I don't have a "photographic" memory or anything like that; the amount of time I've spent searching my apartment for my phone disproves that idea.

While many people, upon learning that, for example, *oviparous* is an adjective meaning "egg-laying," will quite sensibly forget it almost immediately, I will probably remember it, and without any particular effort. I've never trained my memory or anything like that. I did try once or twice, but that was just one of the many self-improvement attempts to run up against my ADD and get abandoned, sometimes in a matter of minutes. No, I simply got a lucky roll of the genetic dice on that front.

Another factor, of course, is my privilege. While I didn't grow up in the most emotionally supportive environment, it was far above the historic average in most other ways: I never went a day without food or a night without shelter, I've never had an invading army occupy my home. And I had parents who were knowledgeable themselves, and who believed in the value of knowledge as its own reward.

Moreover, I am white, and until well into adulthood I was perceived as male. Had that not been the case, then, thanks to the prejudices that permeate our society, my intelligence would have been seen as surprising at best, and threatening at worst, which undoubtedly would have hindered my intellectual development. But because of my demographic characteristics, I was never discouraged from acquiring knowledge. (Well, almost never; I was strongly discouraged from acquiring any knowledge whatsoever about human sexuality—an attempt at censorship that had, shall we say, mixed results.)

My other general approach is to dispute the premise of the question. In other words, denying that I'm even "so smart" to begin with. After all, being able to name all the monarchs from the House of Stuart[6] is a pretty narrow definition of "smart," don't you think? There are many types of intelligence, and the one I have is hardly the most useful. God knows, for most of my life I would have happily traded the type of intelligence that can name the decisive battles of the Thirty Years' War for the type of intelligence that can determine whether or not a girl is flirting with me.

My financial and time management skills are abysmal, and even in the realm of "pure intelligence," I have plenty of shortcomings; I'm terrible at chess, and higher math is just as bewildering to me as to most people. Being on *Jeopardy!* has given me a whole new set of avenues for this approach, because being good at *Jeopardy!* requires different skills than just being smart, even in the narrow sense of remembering who negotiated the Missouri Compromise.

Jeopardy!, like sports, is an attempt to measure a natural talent via an unnatural competition. So you don't just need to "know stuff," you need to know the right kind of stuff. During my run, *Jeopardy!* aired a tournament for college professors, who are essentially *professional* knowers of stuff, yet their collective *Jeopardy!* performance was not particularly elite. *Jeopardy!* rewards breadth of knowledge, not depth, and as such rewards the combination of knowledge and laziness that's been my hallmark from childhood. Buzzer timing is also critical. For whatever reason,[7] it turns out that I'm good at beating my opponents to the buzzer when we all know the answer, and that has made me successful in a way that a

---

6   Or whatever.

7   I'm genuinely not sure. Marching band experience? The hours I spent playing Defender on our Atari 2600?

more knowledgeable person with worse buzzer timing might not have been.

But there's one other skill that *Jeopardy!* requires, and it's the one that failed me in those early days after *Jeopardy!*, when the deflection and self-deprecation I'd always relied on started to feel hollow and dishonest. *Jeopardy!*'s central gimmick ("The contestants give answers, not questions!") is a remnant of the quiz show scandals of the 1950s, and for a long time I thought it was kind of embarrassingly outdated. But I've come to see that this gimmick (if perhaps inadvertently) teaches an underrated skill, which is understanding what you're being asked.

The gimmick of the show forces a weird kind of syntax on the clues such that, oftentimes, you have to untangle the question before you can even begin to find the answer. A clue might read "M is for moonlight, as in 'Moonlight Feels Right' by Starbuck, as well as this xylophone with an unexpected—and amazing!—solo."

That requires some deciphering to realize that the actual question being asked is "What type of xylophone is used in the song 'Moonlight Feels Right'?" or really, "What type of xylophone starts with *m*?" That skill, to cut through imprecise and convoluted language to recognize the real question, is the only *Jeopardy!* skill that's useful in daily life. So it's ironic that it was failing me in life at the same time that it was winning me money on TV.

See, it is still the case at times that, when a person asks me, "How did you get so smart?," it has the same subtext as it did in childhood, meaning "So what, you think you're better than me?" And the approaches I describe above assume that that is always the real question. But the reality is, that's hardly ever true these days. Mostly what people are really asking is "How can I get smarter?" And for that question, my go-to answers are not just wrong, but borderline hostile. Telling you, "I'm just lucky" implies that you can never be as smart as I am, since you weren't born with my gifts. Telling you, "I'm not really that smart" implies that you're delusional, that the intelligence you attribute to me, and aspire to attain for yourself, doesn't even exist.

Deep down, I don't even believe myself anymore when I offer those deflections. They're true in their way, but they're hardly the entire truth. The fact is, I do know how I got so smart, and what it took. The real answer to "How did you get so smart?" is simply this: I wanted to. And if you want to, you can do it, too!

For example: one response I gave on *Jeopardy!*, which my friends found particularly impressive, required knowing the meaning of that word I mentioned earlier, *oviparous*. How could I know such a random fact? Well, if that definition was just a bare, unadorned fact, then it wouldn't interest me, and thus I wouldn't remember it. But in reality, there are no "bare" facts. Everything is a strand in the infinite tapestry of possible knowledge, and I want to explore as much of that tapestry as I can, in the pleasant knowledge that I will never, ever be done. So: *oviparous* is an adjective for animals that lay eggs. I know that *ovi-* means "eggs," related to *ovum*, so *-parous* might mean something like "giving birth." And aha! I remember that an old-fashioned word for giving birth is *parturition*, so that *par-* is probably the same root, and now I have two weird vocabulary words connected in my mind, each of them much more likely to be recalled.

And it goes beyond that: wherever I saw the word *oviparous* originally, it was almost certainly in an at least somewhat academic context. So this is another reminder that scientists have a tendency to prefer obscure words like *oviparous* to straightforward words like *egg-laying*. Why is that? Well, in part, it's a reflection of the fact that, in Western society, academia grew from two sources: the Catholic Church, which wrote in Latin, and the surviving work of Greek philosophers, who wrote in Greek.[8] That isn't in itself a good reason to know those languages, but baked into our academic tradition is a belief that advanced education is for the elite. Having sufficient time to learn one dead language, let alone two, is a pretty surefire sign that you aren't plowing your fields from dawn to dusk in the desperate hope of having enough to eat, and thus must have some relatively high place in society.[9]

The ego hit of that status-signaling is quite sufficient to explain the prevalence of Greek and Latin jargon in academia, and indeed, the tendency toward jargon in any specialized field. So now not only have I acquired the knowledge of the definition of *oviparous*[10] but I've also gained greater insight into how our society organizes itself, and the motivations (and thus implicit biases) that drive scientists. And of course any of these threads will lead in their turn to an ever-increasing array of further threads, on and on indefinitely.

Following those threads is a richly rewarding experience that will not just make you better at *Jeopardy!*, but better at living in society. Better at understanding what is going on around you, and why, and what might happen next, and how you might prepare for it. It will make it harder for you to be scammed. It's no coincidence that extremist talk show hosts promote anti-intellectualism on their shows, and then promote bullshit vitamin supplements and reverse mortgage scams during the commercials. If your

---

8   As you might expect.

9   I suspect we value tradition and classics because we can't help but notice that all the people around us are just as clueless and bumbling as we are, but maybe people in the past actually knew what they were doing. So we follow the patterns they laid down and take a break from having our own imperfect brains be responsible for everything.

10   Which isn't all that useful, tbh.

knowledge is limited, you're an easy mark for people who wish you harm. Knowledge is a shield and a sword, a joy and a duty, and while you may never remember things quite as easily as I do, or win a bunch of games on *Jeopardy!*, if you have the desire, not just to know but to understand, then you will grow more and more powerful every day, and nobody will be able to stop you.

Plus, it's just fun to say. *Oviparous.* Very satisfying.

CHAPTER TWO

When Did You Know You Were Trans?

When I started to question my gender identity, I had to question everything. The entire story of me, which I'd been writing my whole life, now turned out to be severely flawed, or at least to have had an extremely unreliable narrator. And so my story was going to need some serious retconning. It was a *Crisis on Infinite Amys*. So many random plot details turned out to have had a hidden significance all along. And the question I found most pressing was: Had I known this plot twist was coming the whole time? Trans people are supposed to, after all. But had I? After all, if you had asked me, at any point in the first thirty years of my life,[1] whether I was a trans woman, I would have said no. Some people *had* asked me that, in fact, and I had denied it.[2] And yet, when I look back, the evidence was right in front of me all along. When did I know? I'm not sure. But I can see a lot of clues, times in my life when I might have known . . .

---

1 Meaning, not coincidentally, the years I spent almost entirely in and around Dayton, Ohio.

2 More often they asked me if I was gay (i.e., attracted to men), which I also denied, but that's a whole different chapter.

I might have known in third grade. In second grade I loved
the American Girl books. Although, I say American Girl, but I
would almost always just reread the Molly books. Yes, Samantha
and Kirsten were fine, and they had that same quality of being
part of a larger story, while still having relatable age-appropriate
problems and stakes, but I was all about Molly McIntire.

Molly lives in the Midwest and her dad's off fighting in World
War II, and she has her two girlfriends.[3] She tries hard to be
well-behaved, but sometimes circumstances conspire against her,
and she has this beautiful dress that she wants to wear for the patri-
otism pageant that they're doing on the Fourth of July. She wears
glasses and sweaters, and as an adult I now see that she's strikingly
. . . plain[4], but back then I thought she was pretty. I just liked her.
She was kind of like Ramona, except that also she was helping win
World War II. And I appreciated that she was born in the 1930s,
making her about my grandmother's age. Close enough to my
time to be familiar, distant enough to be mysterious.

While I liked the books, I already knew not to express any
interest in the dolls. Dolls were not for me, I had learned. Every
Christmas, my mom would hand-make a doll for each of my
female cousins, in the traditional costume of a different people
each year.[5] I'd never thought much about it; dolls didn't appeal to me, in any case. I had my stuffed animals—Outdoors

---

3   I only had one girlfriend, Elizabeth Jentleson, but she was a really good
friend.

4   I feel bad saying that about a fictional ten-year-old, but, like, you know what
I mean. I wouldn't say it to her *face*.

5   One year the costume would be from India, another year it might be from
Nigeria, or Sweden, or whatever country had caught her fancy. I would help by
using a chopstick to shove the synthetic batting into the arms and legs she had
sewn.

Bear,[6] mainly, with a rotating cast of supporting characters. But they were for sleeping. Otherwise, when it came to personifiable toys, I didn't see the point. I liked Transformers, or most of the He-Man toys, the ones that had parts you could play around with, like cycling Man-E-Faces through his multiple faces, or extending Mekaneck's . . . neck.[7] Beyond that, I didn't get why I should pretend a toy was a person. When I wanted to think about other people's perspectives, to keep developing my theory of mind, I preferred books.

And apart from that, I'd gotten a glimpse at the price of an American Girl doll at some point, and it was one of those numbers that I knew there was no point even discussing with my parents. So with all that, I didn't feel particularly deprived of the dolls. But neither did I feel that there was any reason not to read the books, and thrill as Molly once again nailed her audition to play Miss Victory. But one day, not long after we started third grade, a classmate saw me reading one of the books and said, "Those are for girls. Boys aren't supposed to read those."

"Oh! Okay!" I said, and I stopped reading them. Just like that. I deprived myself of a joy in my life, because it was a joy only girls were allowed to feel. Life already seemed to be full of joys that one wasn't allowed to feel, so adding one more to the pile came pretty naturally. Nonetheless, that's an early time that I could have known. And even before that, the evidence was starting to add up . . .

I might have known as soon as I started being able to dress

---

6   This being the time when Care Bears were all the rage, as well as the time when my family was pinched enough that Care Bears represented an unaffordable luxury, my mom made me a knockoff Care Bear with a tree on his belly and called him Outdoors Bear. He is with me in my bedroom as I write this.

7   Look, I didn't name them.

myself. Don't get me wrong, I never hesitated to abstain from activities once I learned that they were only permitted for girls, but you better believe that I envied them, and not just because they got to Meet Molly. And what I envied most was that girls had so many more choices about how they looked. Boys could only wear a few types of clothes, and only get one kind of haircut. How amazing it would be to have options! To have your hair long or short, with bangs or without, and then whenever you wanted a change you could put it up into a bun or a ponytail or pigtails or who even knows what! And all the different pretty clothes they could wear, are you kidding me? And don't even get me started on accessories![8] Still, the thought of actually having any of that for myself never crossed my mind. But that didn't mean I was happy about it.

I was grateful that the school at Corpus Christi Parish required a uniform. The clothes I wore at school weren't my choice, I didn't have to pick them out. Deciding what to wear outside of school was always stressful for me. How could I decide when every choice I had was ugly? This dilemma resulted in some epically atrocious decisions. Whenever I see a picture of myself from my teens and twenties, I look as though my clothes had been picked out for me by someone whose dog I'd run over. A grayish T-shirt with reflective orange stripes and the logo of a construction company I had no connection to. A solid purple sweatshirt, with some Asian characters printed on a yellow circle on the chest.[9] All of it ill-

---

8   And manicures! Though I do struggle with them. I love having a pretty color on my nails, but keeping a manicure requires many skills I don't possess: Sitting still. Planning. Not constantly fidgeting with the metal closure on my purse.

9   With, I am sorry to have to report, matching purple sweatpants.

fitting, too big, and cut for a different body. Shorts of a neither-here-nor-there length that offended the eye of God and man. I can't believe my parents let me go out looking like that. But then, they weren't fashion plates themselves. For much of my child-hood, we all shopped exclusively at thrift stores. And take it from me, a thrift store in Dayton, Ohio, circa 1992 will not yield many gems, fashion-wise.

I knew I looked like shit, too.[10] But I didn't see any alternative, any possible way of being happy with how I looked. As always, I dealt with it by committing to the bit. These inexplicable choices were my style, dammit. You could mock my odd clothes all you wanted, but that just meant I saw qualities in them that you didn't. My primary motivation in life back then was to be mocked as little as possible. I could never dress right, I could never be cool, I could never be pretty (*wait, I didn't think that, being pretty is gross, shove that down*), I could never have the right not to be mocked.

But people did think I was smart. They'd always be open to believing that I knew something they didn't, like who won the Russo-Japanese War, or how to do algebra. So as soon as I realized that clothes were fair targets for mockery, I focused on how to deflect it when it came. I needed to convince people I knew what I was doing, which luckily enough was one of my biggest talents. I could convince the potential bullies of the world that I had a reason for my clothing choices, but only if I first convinced myself. So I did.

My ugly clothes were a rebellion, a rebellion that only I knew about and whose purpose was unclear. I liked that purple sweatshirt! It was always one of the first ones I picked to wear. It

---

10   I was closeted, not blind.

epitomized my style. It was an unusual style, and not exactly aspi-
rational, but it became an actual style. My mom found a sweatshirt
at a thrift store; it was a grayish sweatshirt that didn't fit me right,
and it had the Harvard logo printed on it, and the word "Harvard"
printed above the logo, in a ridiculous Old English script and,
bizarrely, in a fluorescent shade of pink. She thought it was something I'd like. And it was, in the sense that I was always up for new
ways to practice hating how I looked.

I could have known then that I wasn't a boy, given the fact
that I hated every article of boys' clothing that I had ever seen or
even hypothesized, and I chose my outfits as part of an elaborate
prank I was playing on myself. But I still didn't know, and anyway
I was about to face a much bigger dilemma. The curse that had
been foretold for me since my birth was finally coming to pass,
and the clothes problem would pale in comparison. Puberty had
arrived . . .

I don't mean to alarm you, but hitting puberty when you *already* hate your appearance is not the most pleasant experience. A
lot of the changes that were taking place were ones that I'd been
dreading for a while, primarily that I'd be growing hair seemingly
everywhere, all sorts of places, places where I could not imagine
it serving any purpose either functional or aesthetic.[11] But what I
hated most was hearing my voice change.

I'd been singing in the church choir and in musicals for years,
and while I wasn't *that* good,[12] I enjoyed it, and I knew I could
at least carry a tune.[13] I was an alto. I would stand with the other
altos during rehearsals for the musical *Scrooge* or whatever, and we

---

11   I mean, my back? My *back?!*

12   As my mom pointed out to me after my *Oliver!* audition, among other occasions.

13   Which is more than my mom could say. Not that I'm bitter.

would chat with each other, about a tricky passage in the score, about the soprano who took every opportunity to show off her range,[14] the unjust way in which altos never got the interesting parts, whatever.

Or when my mom and I went to a sing-along for Handel's oratorio *Messiah*, as we did every December, and we would share a music book and sing in the alto section together, and I was at least better than she was, and there were so many people singing that you could sing as loud as you wanted and still feel anonymous. And maybe some of the old ladies standing around you would say something to your mom between songs, about how they enjoyed hearing her little boy singing out so strong.

But then my voice changed, and that was it for singing. I assumed I still *could* sing, but I couldn't sing alto. What was I supposed to do? Sing tenor? Or, God forbid, *baritone*? Go stand over there, with all the boys who were so weird, so socially hopeless, as to actually be okay with singing in public? No thank you! I didn't stop doing musicals entirely; I did some in high school because the Drama Club was my friend group, and they expected me there. But I couldn't enjoy them in the same way. I could never imagine myself stealing the show with an incredible musical solo, as I had once or twice before. That might have clued me in that this whole "boy" thing was not for me. But it didn't. Singing became just another pleasure to toss on the trash pile with *Molly Saves the Day . . .*

I might have known from my time in the theater. Specifically, from my time playing Algernon in *The Importance of Being Earnest*, which I did off and on for a few years, years that more or less coincided with my pubescence. I'd already learned, through

---

14   There's always one.

theater, that it was okay to be a gay man. I'd been raised to think otherwise, but theater disabused me of that notion.[15] It happened during rehearsals for my third play, a production of *Oliver!*,[16] when I was about thirteen years old. Some of us kids were waiting for our parents to pick us up: me, the boy playing Oliver, and one of the girls who was in the chorus with me.[17]

For whatever reason, the subject of Elton John came up in conversation. And the boy playing Oliver (remember, this was in Dayton, in approximately 1993) said something along the lines of "Oh, I don't like Elton John, he sings songs to his *boyfriend*." And the girl, who by the way was cool as hell, a couple years older than either of us, and taller and stronger—this girl grabbed him by the collar and shoved him up against the wall, like Joe Pesci in a Scorsese movie, and said, "This is *theater*. We don't *say* things like that." It was one of the coolest moments I've ever seen in person.

And that was the end of the story for me, as far as accepting gay people. I had to choose, and I chose the cool butch girl over the weedy blond boy who'd beaten me out for the lead role and

---

15   Getting things wrong is how we learn. One time we had to write opinion essays in English class. I did one about reverse sexism, and the girls in the class roasted me over it, and rightly so. It was a bad argument. Every teenager thinks, *Oh my God, I see all the problems that nobody else has seen before me!* That was a day I thought, *Oh shit, I need to backtrack and rethink this.*

16   The one for which my mom had correctly predicted I couldn't sing well enough to get the lead role.

17   No community theater production can survive if they only allow boy parts to be played by boys; a substantial number of the "boys" in the orphanage/organized crime gang that constitute the background of *Oliver!* are virtually always actually girls. Sometimes they wear their hair tucked up under newsboy caps; other times they're just anachronistically female.

also always somehow seemed to be on the verge of calling his parents to come to his defense.

A couple of years later, I was cast as Algernon. I fell into the effeminacy of the role with an eagerness that might as well have been a flashing neon sign saying "AMY IS A WOMAN" that I'm shocked I didn't see. Thanks to my "This is *theater*" epiphany a few years earlier, it didn't bother me that Algernon was, by critical consensus, the gayest gay who ever gayed. Because whatever the deal with gay people was, they were at least allowed to be in theatrical productions. Even my parents couldn't deny that; as Catholic as they were, they were also intellectuals, and the idea that Algernon *wasn't* intended to be understood as a melodramatic swishy fag could only be maintained by the sort of people who didn't listen to NPR, and my parents were *not* going to lower themselves to that level.

The play is fun and challenging. Every line is a hundred words long, and as Algernon you have to deliver many of them while eating cucumber sandwiches. But more of a gift than that: it was a chance for me to sway my hips the way I thought they should sway, to gesture with my hands when I talked, to end way more sentences on uptalk, to hold a glass with the fewest fingers necessary. Not only was I allowed to violate practically every rule in the Boy Code,[18] I was actually *encouraged* to do so. I was rewarded for it with the praise I craved.

I'll never forget being backstage before a production we were putting on in a venue that, while I can remember nothing else about it, was definitely *not* originally intended as a performance space. I was feeling some stage fright,[19] feeling out of touch with my character, worrying about my lines. And my cousin/best friend/castmate Jacob, who would enter a few minutes after me,[20] saw me stressing out, and calmed me down by saying, "Hey, okay, hold up. Take a breath. Now: let me see you do your Algernon walk. You know that always makes you happy."

And he was right. I perked up, thrust my shoulders back and my nonexistent tits out, and sashayed my way back and forth a few times in whatever liminal space it was where we awaited the curtain that day. Letting my body flow how it wanted, letting it show how it felt, that always made me happy. Would it have been nice to feel that happy in any other context than a poorly attended youth

---

18   Which is ordinarily enforced quite strictly.

19   "If you don't have stage fright before you go on, that's a really bad sign," was a thing we actors would say to each other. It was a usefully unfalsifiable aphorism, because if you didn't have stage fright you wouldn't need a reassuring aphorism in the first place.

20   He was playing the first-act servant that time (Merriman, I think?). In other productions he played the cleric, who I believe was named Chasuble.

theater production of an Oscar Wilde play? Sure! But that was the only context that I'd ever heard of, or really even imagined, where I could sway my hips without the imminent threat of physical violence. I wasn't a woman, I was just a boy who really, really, *really* liked playing Algernon. And who liked doing elaborate hair and makeup, donning a silk dressing gown or some other flamboyant Victorian outfit, and swanning about, talking at length about feelings while holding a small, elegant glass of sherry.[21] But that didn't mean I was a woman. Right?

But I also was pretty clearly trans during adolescence, even on those occasions when I was *not* portraying Algernon in *The Importance of Being Earnest*.[22] My dysphoria, my alienation from my own body, was far more intense once my body started betraying me, going along with the boys, doing what they did, growing hair and producing semen and so on. Did I know I was trans? I don't know, but I was certain enough that *something* was wrong that I never really dated anyone until I was twenty-five. There were a handful of fumbling attempts: a homecoming dance with Nikki Jomantas at which I refused to dance,[23] a self-sabotaging love poem to a crush who was unavailable and wouldn't have responded to that kind of gesture anyway.

Every couple of years I'd force myself to ask someone out, go on a single date that would be roughly 75 percent awkward pauses, and then crawl back into my hole like Punxsutawney Phil seeing his shadow. Winter wasn't over yet, and neither was my

---

21   Or anyway, something we all agreed to *pretend* was sherry.
22   And while I spent way more of my adolescence portraying Algernon in *The Importance of Being Earnest* than you or anyone you've ever met, it was still a fairly small percentage overall. In a given day, I was never able to spend more than a couple hours, max, portraying Algernon in *The Importance of Being Earnest*.
23   Sorry, Nikki!

virginity. The obstacle I couldn't surmount was that I couldn't imagine *anyone* wanting to date me.[24]

My dysphoria wasn't the only reason for that, but it was the main one. I hated my body, hated the way it looked, the shape it had, the texture of its skin, everything. If I liked a girl, then how could I justify asking her to look at or, God forbid, touch this droopy awkward mess I inhabited? Girls deserved better than me. They deserved someone who knew how to play by boy rules, someone who could pass a test in the only subject I always failed.[25] I was good at everything except having a body. And, of course, as us savvy kids who sat at the grown-up table knew, you weren't *supposed* to be good at having a body. It was a test you were supposed to fail, and that, at least, I could manage.

And so that's where things were, forever. As far as I was aware. My body could only ever be a source of shame and weakness and embarrassment. I developed elaborate protocols for interacting with it, like a scientist working with a deadly pathogen in the lab. As often as I needed to, I would put on the hazmat suit and do some maintenance: get a haircut, buy a new pair of black Converse Chuck Taylor All Star high-tops, allow myself to be photographed at a family gathering or work event, shave, fill in my height/weight/sex/hair color/eye color on a form, create a social media profile. Even when I met my first girlfriend, Kelly, who was to become my first wife, nothing changed. She was raised in the same environment as me and had many of the same issues; our running joke was "When will we be able to give up on this physical plane and become beings of pure energy?"

---

24   This despite the fact that multiple women over the years said things to me like "You're cute," "I'd like to date you," "I love you," and so on. I assumed they were speaking metaphorically.

25   Well, that and penmanship.

Our bodies seemed to like each other, and we both enjoyed that,[26] but it was as if we'd both owned pets that turned out to get along. It was *good*, but it didn't have anything in particular to do with us. I'd achieved the best outcome I could conceive of: I had someone who would allow my body to do its body things, such as enjoy having slick tight pressure applied rhythmically to its penis, or be calmed by the sensation of being held, or cry, all that weird inexplicable body shit; who would indeed join me in all of those activities as needed, but would always remember that they weren't real. They were fake, shameful body things that we did so that afterward our bodies could get back to smoking weed and talking about *Downton Abbey*.

I'd found somebody who understood my relationship to my body, and together we could pilot the physical forms we'd been saddled with through the rest of our lives. And they all lived happily ever after.

The End.

Except, one night in 2011, I was driving home from rehearsal. I was playing Francis Flute in *A Midsummer Night's Dream*, the member of the Rude Mechanicals who is assigned the lead female role of Thisbe in the play-within-the-play they perform at the end to the condescending appreciation of the nobility in the play.[27] Which meant that my job, my duty, the commitment I had made to my castmates, required me on a nightly basis to put on a dress, and a long-haired wig, and some makeup that was even showier

---

26   We *really* enjoyed that.

27   And of the audience members, all of whom get to feel like the nobles or the peasants, depending on their inclination.

than the standard theatrical makeup,[28] and flitter and twitter and curtsy and so forth. And as I navigated around the construction on I-880,[29] a stray thought crossed my mind.

What if I just wore a dress, like, all the time?

What if I wore it out at, like, a bar, and did all the swishing and swaying I'd done as Algernon, as Francis Flute, just because I wanted to? What if . . . What if some guy walked up to me, and said, "Hi! I'm" (I don't know) "Matt" (or whatever). "What's your name?"

And what if I lifted my hand up to this (apparently tall) man's hand, who was in this thought reaching out to take my hand, and what if I blushed and looked down and away, and said, "My name's Jenny."

Fireworks. Explosions. Cats and dogs living together. Stock footage of large crowds of people cheering. Champagne bottles popping.

Whoa. Okay. That was a surprisingly intense feeling, and I didn't know *what* to do about it. This would have been a fantastic time to realize I was trans, but nope, that still didn't seem like a possibility. Because trans people *know*, right? That's the story? They knew from the moment they were born that they were in the wrong body?

When my cousin came out as a trans man in adulthood, that made perfect sense to me. I had witnessed my aunt and uncle have

---

28   Which is *extreme*, if you see it close up.

29   I've been in California for fourteen years, and I-880 has never *not* been under construction.

epic fights with them for refusing to wear a dress, from when they were five or six. That was a *real* trans person. I'd never demanded to wear a dress when I was a kid, so I must not be trans. You can't suddenly be trans in your thirties! So what was I? Well, I was what I'd always been: weird and shameful. I had to be. If I was trans, then I might feel good about myself. I might feel good about my body. Like some kind of sick pervert.

But still, that feeling couldn't be ignored. Once the idea entered my head, it didn't want to leave. When HBO aired its *Real Sex* documentary series, I watched, riveted, as two beautiful women made out in a bathtub only to stand up and reveal that they had penises. That hit me in a way that I didn't understand,[30] and I began seeking out that genre of porn. I found it confusing, because in society's eyes, you were gay if you liked that stuff.[31] I wasn't attracted to men, never looked at gay porn, so I didn't understand what it meant that I got off to *that*. I felt bad about all of it, but it didn't stop me from doing it.[32] It was just one more reason to feel residual Catholic guilt.

Then I found a genre called "sissy hypno." The idea is that it's hypnotizing you into being a sissy, a man who cross-dresses and gets topped by dominant men. I watched a lot of video clips of what was usually just straight porn, focusing in on the woman's face, overdubbed with lines like "This is what you want to be doing: You want to be dominated by men. You want to suck dick. You want to wear little silly pink things and you want to shave off all your body hair!" And I thought, *You know, I really do want all that!*

My incognito-mode browsing branched out. No longer just porn, but now some new tentative searches: "crossdresser advice,"

30　Still!

31　"Shemales," as they were so tastefully referred to.

32　Of course porn made me feel bad. Wasn't that the point?

"crossdresser forum," "crossdressing tips," "crossdressers Bay Area," "crossdresser tucking detailed explanation."[33]

I found a discreet "social club" for cross-dressers and trans people down in San Jose that also had a shop where they sold necessities: clothes, shoes, wigs, breast forms. I went there after *Midsummer* rehearsal one day, and I bought the thing that fascinated me the most: a pair of black patent pumps. The owner gave me a lesson on how to walk in heels (shoulders back, weight forward over your toes, on each step put one foot directly in front of the other), and then I went home, told my wife that rehearsal had run late, and hid my new prize possession in the back of a closet.

When I was home alone, I would put on my heels, along with one of the wigs Kelly had bought to use for her sketch comedy group,[34] and practice my walk, careful to step only on rugs in case our downstairs neighbor would hear, feeling the sway of my hips and the hair brushing my shoulders. I would see myself in the mirror, and usually I would be repulsed by the pathetic embarrassing man I saw . . . but sometimes I would also just manage to catch a glimpse of myself as the woman I didn't yet know I was, and that feeling was intoxicating, every time. Confetti, streamers, disco ball strobe lights, time-lapse footage of a flower growing from a seedling.

So I had to keep escalating. With the play over, I had no excuse to be down in San Jose, so I couldn't return to the "social club," although I kept the business card they had given me in my wallet, and would take it out sometimes to look at the tagline, written in pink: "Where *everyone's* a lady!" But, of course, I was at Target all

---

33 *Tucking:* A way of arranging your genitals so as not to show a telltale bulge in tight-fitting panties. I never did figure that one out, tbh.
34 Femikaze, a really fun all-female group I loved hanging out with, and helped write for, although I did so secretly, since, after all, I was a man.

the damn time. With the women's wear section right up front by the door, calling to me as soon as I walked in. Eventually I gave in to the temptation, strolling "casually" through the women's section, then when I didn't think anyone was looking, grabbing something (a pair of panties, a skirt, some denim shorts cut way higher than I'd ever imagined before), and shoving it underneath whatever else was in my basket.

Even as I was doing it, I couldn't understand why I was being so secretive. People weren't at Target to monitor me. They didn't care what I bought. And even if they did, I was a married man, had a wedding band. They would just assume I was buying this clothing for my wife. But I couldn't help it; when I was checking out I would avoid all eye contact with the cashier, my heart pounding, feeling my cheeks blush, fighting the urge to blurt out, "It's not for me!" And then I would walk to the car, drive home, and add it to my stash. It was in my sock drawer, and more and more often, as soon as I knew I was alone in the apartment, I would open it up, dig through the mass of indistinguishable white men's crew-cut socks that the drawer officially contained, and reveal a burst of color and satin and beauty. *Wow*, I would think, *I sure do love cross-dressing! God, I wish I was just trans, it would be so much simpler. Oh, well.*

As I began to realize that it didn't seem like I'd be abandoning this new hobby of mine any time soon, I decided I had to tell Kelly before she stumbled across the ever-increasing amount of evidence. I hadn't been happy about keeping it a secret anyway, but at the beginning I hadn't even considered doing otherwise. A man wearing women's clothes *has* to be a secret, right? It just seemed inevitable, like gravity, or Beyoncé. So, despite trusting Kelly implicitly, I was still a little tense when, one night as we lay in bed, I stumbled my way through an explanation of my . . . newfound interest.

Some small part of me expected her to be horrified, to run screaming into the night, to reject me, and cast me into outer darkness. But she didn't. It's not something she would ever have done. She was supportive of what I wanted for myself, but it seemed clear to me that this was not a part of my life she had any interest in sharing. The idea of her husband in a dress did not hold any appeal, how could it? She had no problem with me indulging my cross-dressing desires when I was home by myself, or at the local fetish-themed coffeehouse's "Sissy Nights."[35] She just didn't want a cross-dressing husband to be a major part of her life, which seemed like a reasonable compromise to me. And so everything was settled. Like so many other men on the cross-dressing forum I occasionally posted on,[36] I would indulge myself from time to time as needed, when I had the place to myself. Not in secret anymore, not exactly, but not where Kelly would have to see it; kind of like the way my dad would go down to the basement and paint historically accurate uniforms onto miniature Napoleonic-era soldiers. And that's how life would go until I died, or Kelly did, in which case I'd probably kill myself anyway. So that was that then, everything was finally figured out.

The End.

On October 2, 2016, I helped Kelly carry her last box of things down to the car, feeling stunned, numb, confused, and hollow. I went back up to the apartment that was now apparently "mine" and not "ours," grabbed a piece of paper from the printer cart, and wrote:

---

35   Wait, are you saying your town *doesn't* have a fetish-themed coffeehouse?

36   Under the username *MaybeMichelle?* Question mark included.

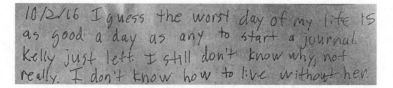

And then a bunch of stuff that's none of your business. Except, at the bottom of that first page, just as an aside, I wrote this:

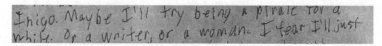

"Maybe I'll try being a pirate for a while. Or a writer, or a woman."

Now did I know I was trans? Still not quite. But my struggle not to know it was clearly getting harder. Some people refer to a person who is trans but doesn't yet know it as an "egg," and that still applied to me. But the egg was cracked, and the crack was widening. There were feathers and a beak sticking out, and I wasn't going to be able to hold on to that shell much longer.

A few months later, I got home from work and, in a routine that had become automatic, went to my bedroom to pick out which feminine clothes I would change into for the evening. As I was debating the merits of each outfit, something occurred to me: if I were to die suddenly, right then and there, I wouldn't get to wear *any* of those outfits. I would be buried in a suit and tie. That's how I would be seen at the wake, that's how I would be remembered. And that last bit of eggshell dropped off.

If I could only wear one outfit for all eternity, I *needed* it to be pretty. And the only way I could be certain of that would be to not *have* any outfits that weren't pretty, *never* to dress in those hideous god-awful boy clothes that I'd hated my entire life. And if I did that, people would need an explanation. And the only explanation that would make sense would be: I am trans. I am a woman. I wear the clothes that were meant for me all along. I was going to have to come out of the closet.

One of the first dresses I bought that I spent some real money on had pockets and it was short and summery. I'm wearing it in one of my favorite pictures from that time that my friend took. We were sitting at this restaurant and I had my back to the window; I'm backlit and just smiling so broadly—a surprise, as I tended not to smile in pictures for most of my life.

So when did I know I was trans? Always. Never. In 1986, or 1993, or 1996, or 2011, or 2016, or 2017, or any and all years in between. But honestly? I've come to realize that I don't much care when I knew I was trans, or even if I know it now. All I care about is that you know. And now you do.

CHAPTER THREE

What's Your Name?

I've been debating whether or not to reveal my deadname[1] in this book. It makes me a little uncomfortable to talk about, brings up a lot of weird emotions. But it's public information. I still get mail with that name on it, and will for at least the next decade or so, and probably for the rest of my life. If anybody wants to use it to troll me on the internet, they can find it.[2]

So let me just say it: I was christened Thomas Ezekiel[3] Schneider. I was Tommy as a kid. Then in first grade I was Thomas be-

---

1 Term of art in the trans community for someone's birth name. Since most (given) names are gender-coded, people who do not identify as the gender they were assigned will often also not identify with the name they were assigned at the same time, and choose a new one for themselves. The word *deadname* seems a bit hyperbolic, but it really isn't.

2 And have. And will continue to. The Google auto-complete suggestions for searching my name can be depressing at times. Not that I'm going to stop Googling myself.

3 One of the Bible's weirder prophets; I've long suspected that it was chosen at least in part by my father in memory of the acid trips he took back in his "misspent youth," as he always called it. Because, dude, the prophet Ezekiel was clearly tripping balls.

cause Mrs. Coleman called everybody by their birth certificate name. So, like, she called Tony Niccolini "Antonio," that sort of thing. I assumed that in second grade I would get to be Tommy again, but when I showed up on the first day of class, there was a new kid there that had also been christened Thomas. His dad had been transferred to the local air force base,[4] and our teacher said that, because it was hard to be the new kid, he would get to choose what name he wanted. He chose Tommy, and I was stuck with Tom. I lowkey resented him for years.

That first day of second grade was, as far as I was aware, the last chance I was ever going to have to pick my own name. I mean, not really, I knew plenty of kids who changed their name. Julie decided to be Julia. Samantha became Sam. Bill became Alex (it was his middle name). Even some of the grown-ups in my life had changed their names; I had two unrelated aunts who were known by different names to the outside world than they were within our family, in both cases because they were named Mary.[5] But the only options I was aware of were Tom, Thomas, and Tommy.[6] Tommy had been stolen from me. Thomas felt like somebody I'd never met. Tom it was.

And then, a few decades later, Tom it wasn't. It became clear to me that Tom was not my name; or at least, that the name Tom would lead people to assume that I was a boy. Now, that was just

---

4   Wright-Patterson Air Force Base. When I was a kid, we were told that it was so important that Dayton was third on the Soviets' list of nuclear targets, after DC and NORAD. I've never attempted to verify that fact, because I want it to be true and it almost certainly isn't.

5   Every Catholic girl in America was named Mary for a few decades. All of them.

6   "Thom" had not yet occurred to me; the first Thom I would know about was even then forming a band called On a Friday with his schoolmates across the Atlantic. A few years later they changed their name to Radiohead.

one of many things that had led people[7] to assume that for my entire life, but many of the others could only be changed (when they could be changed at all) with a significant investment of time, energy, money, makeup, prescription drugs, or extensive surgical intervention. But changing my name, at least in casual conversation among close friends, that was something I could do easily.[8] But while I knew Tom was not the answer . . . What was?

I struggled with it for a long time. I don't know how long it was, objectively, it might have been maybe like three weeks. But it felt long. It felt unending. It's a strange feeling, going through life without a name. One of the first times I'd gone out in semi-public, dressed as a woman, was to a backyard concert at the house of Nora and Ann, this slightly older lesbian couple I knew.[9] At some point during the party, Nora took an opportunity to ask me privately, "I just wanted to check, am I still calling you Tom? Are you still using he/him pronouns?"

"I have no idea," I said. "I don't know what you should call me or what I'm doing or anything. I just . . . I just wanted to wear this outfit. I don't know what's going on with me at all."

And that was it for the time being. She was understanding, didn't demand further explanation. That wasn't true of everyone. One time, my friend Jessie and I were hanging out at our friend's[10] house,[11] chatting about my identity, and he decided that I needed to be pushed. In a very well-intentioned way, he started getting in my face: "You know what? You need to stop debating this, you

---

7   Myself included.

8   "Easily" by the standards of transition shit, which isn't saying much.

9   Hi, Nora and Ann! It's been a while. We should get a drink sometime or something.

10   Our coke dealer.

11   His parents' garage.

know what your name is. You know! You know what it is! Stop
being all like 'Um, well, uh, I don't know,' just decide! Listen, I'm
going to ask you right now: What's your name? Now, what name
popped into your head? Whatever it is, that's the right one. Come
on, you've got to do this now!" It was a bizarre moment, having
somebody attempt to bully me into doing something empowering. In any case, it wasn't the right time.

I did have a name pop into my head, not that I was going to
tell him that. It was a name I'd liked ever since I was a kid, the one
I'd used in my head the first time my egg cracked, the first time I
imagined introducing myself as a woman. *Jenny.* And I know some
trans people who would say, "Well, that's it, that's the name you
gave yourself in childhood, that's your real name and always has
been." But somehow, it didn't feel right. I mean, part of it was that
I had a friend named Jenn, and I didn't want two leading characters in the story of my life to have such similar names. Seemed
confusing. And apart from that: sure, I'd liked the name "Jenny"
when I was a kid, but that was because it was a kid's name. I wasn't
a little girl. I never had been,[12] and I never would be. I needed a
name for my grown-up self.

And, apart from the gender coding, I'd liked the name "Tom"
just fine, after I got over my resentment of that seven-year-old
name thief.[13] It was short and generic, known widely, easy to spell,
and easy to pronounce.[14] People didn't associate anything in particular with it.[15] It felt like with the name "Tom" you could be
any age, do anything for a living. That was what I was looking for

---

12   Or had I? It's confusing.
13   Jk. I never got over it.
14   All of those statements are only true within a particular Anglophone culture, of course.
15   Apart from, you know, testicles or whatever.

in a name. I went through a few rounds of trying out candidates, having Jenn or my therapist try calling me Michelle or Alice or whatever, but none of them felt right.

I don't even remember where or when I came up with "Amy," but that's part of why I know it was the right choice. It came into my brain, and it just fit. It wasn't inspired by anything in particular, it didn't mean anything in particular, any more than Tom ever had. It was just a name, not hard to write, not hard to spell out loud over a bad phone connection to someone with minimal English. Alternative spellings existed, but the A-M-Y version was dominant enough not to require specifying, any more than I'd needed to specify that my name was Tom rather than Thom before.[16] It meant what I wanted it to mean: nothing.

That's something I think people can misunderstand about trans people, trans women in particular. For much of my life, "drag queen" and "trans woman" were understood to mean the same thing,[17] and so like everyone else I assumed that trans-ness was about exoticizing femininity, about wanting to be big and bold and femme and have dramatic hair and makeup and be brash and enticing and intoxicating. That hasn't been completely absent from my experience, especially when I was starting out. When I first came out, it was as if I'd been a kid standing with my face pressed to the window of a candy store for thirty years, and once I was allowed in, I wanted *everything*.

But when that rush settled down, I realized that wasn't what it was about at all. Sure, I liked glitter, sparkles, high heels, bold lips, dramatic eye shadow. But that was just accessorizing. One item of clothing that called to me early on was a shirt I saw at Target: a

---

16   Thom Yorke is someone many people admire, but few seek to emulate.
17   Specifically: "faggot."

shapeless workout/errands/sleep top that said "Messy Bun / Target Run / Get It Done." That was what I wanted, to live my whole life as myself, including the vast majority of it when I wasn't doing anything interesting or girly or exciting. I just wanted to be, well, Amy. That was all.

The final stage of my coming out, the moment that I use to mark my "traniversary," was coming out at work. The reason this was the final stage was not that my coworkers were the people I was most frightened to come out to; I was fortunate enough to be working with a great bunch of people, and though there were one or two engineers who I thought might be a little weird about it,[18] I didn't foresee any major problems.[19] But it still felt weird. It still felt like I was talking about . . . you know . . . *ess-ee-ex* at work. Nonetheless, I needed to do it, I needed to stop having to haul out this ugly, ill-fitting boy costume every day just to do my job.

We had our weekly staff meeting on Friday afternoons, and that meeting was clearly the right occasion for this announcement. For a few weeks I kept deciding that next Friday would be the one, and then finding a reason to wait another week. But then one Friday, instead of having our meeting, we had one of our quarterly[20] "offsites," the equivalent of when your class gets to have a pizza party because they sold the most magazine subscriptions during the annual fundraising drive.[21] Our boss chartered a party bus and we all went up to Sonoma, did some wine tasting.

---

18   They weren't! Everyone was great. We still have a group chat together.

19   I mean, I did, in a sense. I foresaw all sorts of disastrous outcomes about *everything* at this time in my life. I just mean that, when my brain presented me various catastrophic scenarios that might ensue from coming out to my coworkers, I didn't find them particularly plausible.

20   -ish.

21   Everyone did that, right?

I'd decided that the next Friday would be when I would come out, and after a few wine tastings,[22] I pulled my work wife into a corner of a wine cave[23] and came out to her, and told her I'd be coming out to everyone else on Friday, but to keep it to herself until then. So now I had to do it, otherwise she would start feeling uncomfortable about keeping the secret, and I couldn't do that to my girl Candace!

So the day came: June 30, 2017. There was a portion of the meeting where everyone was invited to raise any points for discussion that they wanted, and I raised my hand. I'd rehearsed this moment over and over again for the last few weeks, getting it down word for word. I knew everyone in that meeting, I trusted them all, I didn't see how it could go wrong. Yet I was terrified. I think it was just that, this was it. This was the last chance for it all to come crashing down, for some negative consequence of my transition to turn out to be so bad that I had to just scrap the whole thing and crawl back into my shell.[24] I was trembling as I stood up, and my voice shook, but I forced out the lines I'd been rehearsing. They went something like, "Hey, so, when all of you met me, I introduced myself to you as a man, named Tom. But for the last few months, outside of this office, I've been introducing myself to everyone as a woman, named Amy. And that's how I see myself now. I've been hiding this at work. Well, not completely hiding it."

---

22 Though by this point "tasting" was no longer a plausible description of our activity.

23 Or wine cellar, I think? My memories of all the vineyards after the first one are a bit hazy.

24 The shell was broken into a million pieces and scattered all around the coop by then, but I still thought it could happen. Have I stretched this metaphor too far?

I held up my hand, which was shaking, to show off my manicure. I'd been painting my nails for a few months by that point; when asked about it I would say something like "Oh, you know, I just felt like I wanted to do some weird San Francisco thing. It's fun!" So this was my acknowledgment to my coworkers that my explanation had never made any sense, and we'd all kind of known it. It got the exact sort of response I wanted, a professional work meeting chuckle; not insincere, but professional. I said a few more sentences, partly just around some practical details involving my email address, and something about how, as a software engineer, I know that two different pointers can refer to the same value, and so if people called me Tom I would know what they meant. It was a solid work coming-out speech. Self-deprecating, professional, to the point, didn't waste too much meeting time. The quivering mess of barely concealed emotion with which I delivered it wasn't ideal, but that's why I'd made sure to write myself some good lines, to paper over the weak performance.

And . . . that was that. It was done. Everybody said the right things, and meant them. The meeting ended, and I went back to one of our single-person bathrooms, where, for the last time, I took off my stupid pants and my baggy, ugly T-shirt and my disgusting boxer briefs, that whole mass of rotting scar tissue I'd been lugging around forever. Then I put on the clothes, the *real* clothes that I'd brought in my backpack, put on some makeup (I wasn't great at it, but I did my best), and finished out the workweek. The last chance for disaster had come. Disaster hadn't happened. I was free.

On the one-year anniversary of their hire date, every employee was given a coffee mug with the company logo and their name on it. I'd gotten mine a few months earlier, which bummed me out, but whatever, it was just a coffee mug. A few days after I came out, our office manager walked up to my desk, carrying a box. "Hey, Amy, this just came for you." She pulled out a coffee mug. Just like the one I had. Company logo on one side. And on the other side: my name. Amy. If it was on my coffee mug at work, it had to be real.

CHAPTER FOUR

What Teachers Made a Difference to You?

When I was a senior in high school, my mother was teaching a math class at the University of Dayton, the college I would be attending a year later. Since she hadn't completed her PhD,[1] she usually taught "math for jocks" type classes, for students who hated math but needed to meet the bare minimum to graduate. One day, she wasn't feeling well, and sent me in to monitor an exam she was giving. I was taken aback, but she said it would be fine, all I had to do was sit there and make sure the students weren't cheating.[2]

Now, my mom had been a teacher my whole life, albeit usually part-time, but for some reason I never really pictured her in a classroom. So when I got in and saw the exam questions she had prepared, I was startled. They were great! I don't really

---

1 Due to some absolutely shameful behavior by some people in academia. I don't want to publicly attack anyone based on my secondhand understanding of a decades-old incident, but I think it is safe to say that, for women, graduate-level mathematics was not exactly a welcoming environment in the '60s (or, you know, today).

2 Not *blatantly* cheating, at least.

remember what the test instructions said, but they had character, some sarcastic jokes, a little encouragement. I thought, *Oh
damn, she's fun!* I don't know why I was surprised.[3] She would
talk about the anecdotes she would tell in her class, or assignments she could give, and in retrospect they all sounded like the
type of teaching I enjoyed. The classroom brought out the best
in her.

And, of course, she'd always been a good teacher to me at
home. One of the earliest such times I can remember[4] was when
she did an experiment with me. She had two glasses of water, one
tall and skinny, and the other one short and wide. She poured
water from the wide glass into the skinny glass, and then asked
me if the second glass had more water than the first one. I said
yes, clearly it did, because the water reached much higher in the
skinny glass. She then tried to explain to me that it must be the
same amount of water, since I had seen her pour it from one glass
to the other, but I was still too young to get that. But it always
stuck with me, I think, because not only was she teaching me, but
she was teaching herself, running an experiment to demonstrate
what she'd read about the rate at which children develop an understanding of conservation. I think one of the requirements for
being a great teacher is having a sincere, contagious interest in the
things you're teaching. My mom was the first great teacher that I
had, despite never having me in her classroom.[5] Let me tell you
about some of the others.

---

3 Well, I know partly it's because I was a teenager, and she was my mother, and
I wasn't much inclined to give her the benefit of the doubt on anything.
4 It must have been when I was younger than eleven, according to the research
of Jean Piaget.
5 Thank God.

First Grade: Mrs. Coleman

Mrs. Coleman was a legend at my parish school. She was known as a disciplinarian, but she wasn't, really. She was just a strong, determined personality who made it very clear at all times that she was in charge and we were doing things her way, and that was the end of the discussion. When we were out in the world, on field trips or whatever, she'd carry a bell, and when she rang that bell, all of us would immediately stop whatever we were doing and line up behind her. Over the years, I would happen to see her with her students now and again, and it was still just the same, which I find awe-inspiring, as someone who now understands what chaos agents six-year-olds can be.

She was also the only Black teacher I ever had, in all my years of schooling. And she was passionate about Black history. Martin Luther King Day was the biggest day on the first-grade calendar, with a pageant that we spent more time on, and were more excited about, than the obligatory Christmas pageant the month before. Her classroom had posters lined along the top of all four walls, each with a picture and short bio of a different Black activist or pioneer. It wasn't just MLK and Rosa Parks, but Phillis Wheatley and James McCune Smith and many others. Obviously, as first graders, we weren't able to really understand the complexities and traumas of Black history in America, but she was at least able to deliver the message that Black people could be heroes, could do great things, could be role models to aspire to, whatever your race. I don't want to claim that I came out of first grade an enlightened progressive, but I know that Mrs. Coleman inoculated me at least somewhat from uncritically accepting the racial prejudice that permeates our society. It was a brave thing to do at a school where she was the only Black employee, and I've admired her ever since.

My mother likes to recall overhearing a conversation I had

with my best friend at the time, Elizabeth Jentleson.[6] On the last day of first grade, one of us said something along the lines of "You know what? In kindergarten, everybody said Mrs. Coleman was so scary, like, 'Oh, you better stop acting like that or you're going to have to deal with *Mrs. Coleman*,' but I like her!" The other one said, "Me too. She was a good teacher." My mom quoted us for years. She loved that we had come to appreciate the supposedly scary tyrant Mrs. Coleman.

Eighth Grade: Mr. Reitz

Mr. Reitz was actually my teacher for at least one class from grades four through eight, and if you've ever been around fourth- through eighth-graders,[7] you know that they have *drama*. Mr. Reitz was there to help us through it. He was the only male teacher at the school, but he was also the most sensitive and adept at conflict resolution. When tensions were running high, like if a particular bully was getting out of hand, or a disputed call in a kickball game at recess had escalated to the point of Shawna sitting on home plate with the ball crying while the rest of the class stood there yelling at her,[8] Mr. Reitz would call a "circle meeting." We would

---

6   I say "friend," but, truly, we were in love. We walked back and forth to school together every day, and sat next to each other in class; we drilled two holes into a rubber eraser so we could put it on the border between our desks and use it as a shared pencil holder. We had our own special codes to communicate how much time was left in class, and how excited we were about that. Her parents moved away in the summer between second and third grade, and on the last day before the move we sat silently side by side on our porch swing as our parents wished each other well. We knew if we said we loved each other nobody would believe us.

7   And haven't blocked out the memory to preserve your mental health.

8   Just a random, purely hypothetical scenario, certainly not a real event in which I was one of the ones yelling at the poor girl.

form our desks into a circle, Mr. Reitz included, and talk out whatever the issue was, using "I" statements ("I feel x when you do y, because z").

He loved teaching, and not just about whatever the subject at hand was. He subscribed to the *New York Times*, and every Tuesday he would read the Science section and come in excited to share some new research with us. He had extra little facts for everything we were studying; the one that sticks out in my mind was about how the guns in the Old West were incredibly inaccurate, and you could shoot at someone and end up hitting someone else way off to the side. In retrospect, many of his fun facts were somewhat exaggerated or misremembered,[9] which is maybe not the *best* quality in a teacher, but the important thing was the excitement, the genuine joy he had in being able to share something with us that we didn't know.

On the last day of eighth grade, he asked us how we wanted to spend our last hour, and somebody raised their hand and said, "Can we have a circle meeting?" And for one last time, we got our desks in a circle, and all went around saying how much we'd meant to each other, and how much we'd miss those of us that weren't going to the same high school. Among the things I learned from him[10] was this: if you are completely sincere, and stubbornly persistent, you can eventually get middle schoolers to buy into the corniest shit imaginable.

---

9 Which was mostly harmless, although once he said something about how a needle could get into your bloodstream and it would eventually reach your heart and kill you, and a few months later I got a needle from a cactus stuck in my finger and spent a day convinced that it was my last day on earth. Although, I've developed such a fondness for that anecdote over the years that it all evens out.
10 That were actually true.

Theater: Jean Howat Berry, Lisa Howard-Welch, Fran Pesch

Some of my best teachers never taught me in a class, like the soccer coach in eighth grade who motivated me to actually exercise regularly, outside of soccer practice, which is a gift I've only grown to value more as I've gotten older.[11] Or the leader of our church youth choir, whose love of music, and belief in us, was so powerful that we found ourselves singing quite beautifully without quite knowing how we did it. But the "extracurricular" teachers I remember most were all in theater, and three of them shaped my life profoundly.

Jean Howat Berry ran the summer youth theater camps at the Dayton Playhouse. Now, if any of you are familiar with children's theater programs, you know that they function primarily as day care. But Ms. Berry tried to push us to do real theater, and to understand what made it more than "playing pretend." I'd done my first play because my cousin's college production of *The Music Man* needed town kids, and it was certainly fun.[12] But then that summer I was at camp and Ms. Berry suddenly came in and said, "Come on, I want you to sing for these people," and next thing I knew I'd been cast as Young Patrick in *Mame*, a fairly substantial role. If you were a kid willing to buy into theater, she would give you every opportunity she could. I learned a little of everything: how to light a stage, how to run sound, how to dress a set, how to call a show as a stage manager. No part of the playhouse was

---

11   Do I exercise regularly? Not at the moment. But I know how to, not just in theory but in practice, and comparing notes with other not-particularly-athletic adults, it seems like that makes a real difference. I'm not *good* at getting into a regular exercise routine, but I know that I can, and therefore sometimes I actually have. I ran a 5K once!

12   Any time I got to hang out with grown-ups was fun.

"adults only" when we were with her. And she produced some great[13] theater along the way. Seriously, it blows my mind when I realize that this woman looked at a group of Midwestern kids whose parents wanted a few hours to get some errands done without anyone nagging them for fruit snacks, and decided, *You know what we should have them do? I know! Let's have them perform an experimental German play from Weimar Berlin that explores the way that morality emerges from the underlying economic relationships in society!* And she was right![14] It was *The Good Woman of Setzuan*, and while my part was small, it might be the production I am most proud to have been a part of.

It also blows my mind that anyone would say, *I'm going to devote decades of my life to creating and running a children's Shakespeare company in Dayton, OH*, but by God, Lisa Howard-Welch did exactly that. I hope at some point to find a long enough time, and a compliant enough publisher, to really tell the story of W. Shakespeare & Co. as I experienced it.[15] For now I'll just say this: Lisa taught me that there was nothing wrong with attempting the impossible. It was highly unlikely that a nonprofit youth theater company focused on Shakespeare and other Anglophile-friendly works[16] would take root and become a well-regarded artistic institution, but she decided to try it anyway. Because whether or not that grandiose dream succeeded,[17] the attempt would be a worthwhile outcome on its own. And as far as I'm concerned, it was.

---

13 Yes, yes, grading on a *very* steep curve, but still.

14 In a funny way, Bertolt Brecht turns out to be pretty solid children's theater material. Kids respond well to Marxism.

15 I touch on it in chapter 6, but that is just the tip of the iceberg. I mean, the events surrounding our production of *Love's Labour's Lost* alone could fill novels.

16 Jane Austen adaptations, Oscar Wilde, Tom Stoppard, etc.

17 It didn't, unless I've missed some major developments since I moved away.

W. Shakespeare & Co. was a terrible idea, and I will forever be grateful that she had it.

Drama Club was an extracurricular at my high school,[18] so Ms. Pesch wasn't technically my teacher. But as far as I was concerned, Drama Club was a required class, never mind the fact that it took place outside of school hours and nobody received grades. Some people teaching theater to kids are great because they are so focused on the children themselves, and empowering them to be their best. Not Ms. Pesch. Ms. Pesch was great because she was focused on the *theater*. She was there to put on the best show she could, and the fact that we were children was just one more obstacle she had to overcome.[19] She was an advocate for the audience, not for the cast. She understood that, when the Drama Club put on one of their handful of shows each year,[20] nearly everyone in the audience would be there out of obligation, because they had a friend or relative in the cast and couldn't say, "No, I actually don't want to watch a bunch of teenagers flounder their way through a production of *Guys and Dolls*."

Her notes always emphasized keeping the show moving, not leaving any dead air. *Pace, pace, pace!* she would say. And she was always moving herself—she never dragged, never held back, never

---

18   The idea of getting academic credit for doing theater still seems like an absurd fantasy to me. But then, so does getting paid to go on TV and answer trivia questions. I'm beginning to question the criteria I was taught to use when differentiating fantasy from reality.

19   Producing any kind of theatrical production, whether live or filmed, consists entirely of navigating over and around an infinite series of obstacles. It's like playing Super Mario Bros., except that all the mushrooms and turtles are real people that you might need to work with in the future.

20   A musical in the fall, one in the spring, and a straight play in between, plus some other things sprinkled in: a Night of One-Acts here, a War on Drugs propaganda piece funded by the D.A.R.E. program there.

dawdled. When she was directing, whether she was blocking out a scene in rehearsal, or making a general announcement about the rehearsal schedule, or running an audition, or ripping into us after a disastrous tech run-through, she was never, ever boring.

And she was also never, ever hurtful. All of the above might make her sound like a bit of a tyrant, belittling her students because they don't live up to her delusions of grandeur. But she wasn't that. She never asked us to do things that were beyond our capabilities, but she wouldn't let us do less than we were capable of, either. Not because we owed it to her, or to ourselves, but because we owed it to the audience, those poor parents sitting in the audience, sneaking peeks at the program to figure out when they'd be able to go home and watch TV. She taught me that the work that you do doesn't matter unless it is being done in the service of others, and that doing well isn't enough. You can't feel satisfied unless you know you did your best. Not *the* best, just the best you had in you that day.

Freshman Religion: Ms. Blier

Ms. Blier was the first teacher I had a crush on. Like all my crushes, it was a confusing mix of attraction and desire, of wanting to be her girlfriend and to be her.[21] She was fun and cool and bohemian and let us have class outside sometimes. She was the first liberal Catholic I'd ever encountered, and I was fascinated. Freshman year was the Old Testament, and she taught us a feminist reading of that patriarchal[22] text. I particularly remember her portrayal of Samson as a big dumb idiot, with Delilah as the long-suffering

---

21   For more on my trans identity, see every other part of this book.
22   Literally.

girlfriend exasperatedly keeping him out of trouble. Ms. Blier showed me that a text could be interpreted in different ways, and that when you are being taught something, you don't just have to choose between accepting or rejecting it, you can also choose to make it your own, no matter how authoritative the source claims to be.

I think she liked me, too; certainly I used all the grown-up pleasing skills I had on her, and we had some good banter over the course of the year. She got engaged in the spring, and brought her fiancé in to meet the class. He told us all to treat her well so that he wouldn't have to deal with her being grumpy when she got home. I then raised my hand and told him that he should treat her well for the same reason, and he turned to her and said, "Oh, this must be that Tom[23] Schneider you're always telling me about." I was clearly flustered, and after he left the room Ms. Blier laughingly pointed out that the textbook on my desk had my name printed on it in big letters. I was proud to be teased as if I were a peer and not a student, but a little disappointed that she didn't actually talk about me all the time.

High School Science: Ms. Anderson and Mr. Korzan

Ms. Anderson was my biology teacher, and I'm so glad that she was. Biology was one subject I never enjoyed. So much of what it dealt with was icky and gross, in my opinion.[24] And while I can't say I grew to love biology, I wasn't miserable in biology class, and Ms. Anderson was the reason why. It simply came down to the fact that she loved biology passionately, and the joy she got from it was

---

23 That being the alias I was going by at the time.
24 My standard line was "If God wanted us to know what our insides looked like, he would have given us tiny windows."

infectious. I think any great teacher has to love the subject they teach, and Ms. Anderson was definitely a great teacher.

Mr. Korzan taught me chemistry and physics, and seemed to have a great time doing it. He was consistently one of the least prepared teachers I'd ever had, but he made it work. In the very first week I had him as a chemistry teacher, he had us all do an experiment to combine Chemical A with something (Water? Baking soda? I don't remember) to turn it into Chemical B.[25] Only, afterward, he found out that Chemical B was hazardous, and couldn't be thrown away or washed down the sink, but needed some kind of special disposal. So the entire rest of the school year, there was a tray in the corner of the lab filled with beakers of Chemical B. I never did find out what Chemical B was, or whether he ever figured out how to get rid of it, or just left it there over the summer and hoped someone else would figure it out. It wouldn't have been out of character if he had.

He always wrote our tests at the very last minute, often in the car on his way into school that morning. So when we were taking the test, he'd be taking it himself, at his desk, and would occasionally say things like "Hey, everyone, on question four, that's supposed to be meters per second, not miles per hour," or "Hey, everyone, just skip question eight, it's not actually solvable the way I wrote it." He also made us into characters in his word problems, and if you'd been annoying him, you could expect a question on the next test to say something like "Matt[26] is riding his skateboard at 5 meters per second when he suddenly rolls off the edge of a cliff that's 200 meters tall. How far does he travel vertically before

---

25   That's what he called them, "Chemical A" and "Chemical B." The point wasn't the specific chemical reactions involved (this being the first week), but just to get us familiar with doing things in the lab.

26   For example. But it usually was Matt, tbh.

smashing into the ground, and how fast is he going at the moment of impact?" Making learning fun isn't a simple task, but Mr. Korzan made it seem easy. For better or worse, he's one of my role models for doing things at the last minute, and getting away with it.[27]

College: <Names Redacted>

Unfortunately, I had very few good teachers in college. Some of that was just luck of the draw, and some of it was the nature of computer science programs. The thing is, if you have a PhD in computer science, you can make way more money doing anything else than you can make as a college professor, especially if it's not a top-tier program.[28] So the professors we got were usually rather unimpressive, and at times flat-out terrible.

One in particular stands out: in every class he would read out his slide decks in a monotone, and answer questions by simply repeating whatever he had just said. He would put questions on his exams that had nothing to do with the material we'd been studying, and his grading was haphazard; one time two students turned in identical answers to their tests (he made no effort whatsoever to prevent cheating), and got grades that differed by eight or nine points. He assigned us a huge project that we spent months on, and when we each handed in our thirty- or forty-page results, he sat in front of us in the classroom, flipped through each one for maybe ten seconds, and then assigned it a grade. At the end of the semester, we marched down to the department head en masse to demand that he never teach there again.

There was one professor that was pretty good, although sadly

---

27   Writing a book, for example.
28   Which the University of Dayton's was, uh, not, as far as I could tell.

I've forgotten his name. He taught me one of the only things I took away from college, which was "There are no win-win scenarios in computer science," an insight that applies to most things in life, I've found. He also did something I've always appreciated, which is that when I was about to graduate, he refused to write me a recommendation letter for prospective employers. He said to me, "Look, you're smart, you know the material, but you never did your homework, and in a workplace, you have to actually do the tasks you're assigned." You might think I'd be resentful, but in fact I felt nothing but respect. I knew perfectly well that I was a slacker, that I made a habit of half-assing everything and counting on my natural abilities to avoid any serious consequences. And I felt bad about it; I regretted the opportunities I could have had if I'd worked harder, if I'd maximized my abilities, and I resented all the people in my life who had so consistently allowed me to get away with my bullshit. It was nice that, in the final days of my academic career, someone finally told me that they saw me, they understood what I was doing, and they were disappointed that I wasn't doing better than "good enough."

I also had a disastrous creative writing teacher. I won't name him here,[29] but he's one of the only teachers that to this day I actively resent. He loved white male authors, and hated genre writing. He wanted us to write like the authors he loved, particularly John Updike and, above all, Ernest Hemingway. He wanted terse, emotionless sentences, with characters whose complexities were only revealed through the subtlest hints and word choices, depicting quotidian activities in settings that were either banal or unpleasant. He refused to accept anything that wasn't naturalistic,

---

29   Though I suppose anyone who's deeply familiar with the makeup of the University of Dayton English department at the turn of the century could figure it out.

with a particular contempt for science fiction, which he explained by saying, "Why would I want to read a story about a robot that falls in love with a toaster?"[30]

I came into his class loving writing; had my parents not prohibited it, I'd probably have majored in English. After two semesters with him, I hated writing, and it was some time before my hatred faded. He single-handedly ruined one of my greatest joys, and I'm still a little bitter. But guess what? I'm writing a book now, and violating as many of his rules as I can manage. And it's coming out great. Screw you, Hemingway.

High School English: Mr. Brooks

Which brings me to the best teacher I ever had: Mr. Brooks. He taught me English freshman year, Creative Writing in at least one other year, and European Authors my senior year.[31] He was tall and thin, with bags under his eyes.[32] He was extremely funny, but completely deadpan. Some of his lessons were legendary; the day he would teach prepositions, he would act them out, climbing up on his desk for "over," crouching under the tall chair he sometimes taught from for "under," and out the other side for "through," things like that. He oversaw the annual publication of *In Our Minds*, a collection of art and writing produced by students that year. When it came out, he would go around the classrooms trying to get people to buy it. Part of his sales pitch was "You should buy one before we run out. They're going like hotcakes. And you

30   I'd read that! Wouldn't you?
31   I couldn't take Honors English my senior year due to some kind of scheduling conflict, but I didn't mind, because it gave me one more semester with Mr. Brooks.
32   I would joke that he looked like the ghost of Jacob Marley.

know how hotcakes go. Fast." I know it doesn't seem that way on the page, but I swear, when delivered with his deadpan delivery and pitch-perfect timing, it was one of the funniest things I've ever seen.

And as a writing teacher, he was the complete opposite of my college professor. He wanted us to be familiar with a wide array of forms and styles, and he always supported us in whatever direction we wanted to go with our work. Senior year I'd discovered Kurt Vonnegut and gotten super into his work,[33] so for my next assignment I turned in an essay that was a blatant Vonnegut imitation.[34] Short paragraphs, frequently jumping from one topic to something seemingly unrelated, breaking the fourth wall, using colloquial language. It was far afield from the assignment, but he didn't mind. He saw how much I'd enjoyed it, how much fun I'd had with it, and that was all he wanted.

And that's the thread running through all these memories, connecting all the good teachers[35] I've ever had. A good teacher doesn't just teach their material, they advocate for it. They show you not just what you should know, but why you should care about it, often simply by showing the pleasure they take from knowing it themselves. I'm lucky to have had so many teachers do that for me, and my hope is that I can do the same for you. I hope I'm able to show people the joy I take in learning, in knowledge. If I succeed in that, it's all thanks to the names in this chapter, and I'd be honored to have passed on the gift that they've given to me.

---

33   A smart teenage boy getting really into Vonnegut? I know, shocking!

34   *Hocus Pocus*, to be specific.

35   And there were plenty of other good ones that I didn't mention here, apologies to any of them who read this. I'm grateful to you as well!

CHAPTER FIVE

What's It Like to Be Famous?

I can't speak for you, but if I were reading this book,[1] this is the part I'd be most interested in reading: the part about being famous. My whole life I had wondered: What would it be like, to be one of *those* people? The winners. The people who hit the jackpot, who go from normal life to that other kind, where your name is in headlines and your face is on TV, where you get recognized on the street, have your own fan club and Wikipedia page. How would that feel? No, but really, how would it *actually* feel?

And so I want to warn you right away: I still don't have an answer to that question. I've gotten a lot closer, but this is all still so new. It has only been a year since I found myself living my fantasy, a year that passed like a whirlwind, a whirlwind that picked me up and carried me along with it, twisting and tumbling.

And I've fucking loved it! I'd always wanted it, on some level. After all, that's the whole reason I wanted to know what it felt like! But I hadn't admitted that to myself. Officially, my position was that fame wasn't for me, that I'd be overwhelmed by it. When I

---

1 And if I didn't already know myself personally.

married Kelly, who by majoring in theater had publicly declared her intent to pursue some form of stardom, I thought I'd found the perfect solution. Kelly would be the star, and I would bask in the warmth of her stardom, without having to risk exposing myself to the glare of public scrutiny. My public image would consist of blushing modestly while Kelly thanked me in acceptance speeches for various awards, and that would be enough for me.

Well, that would have been enough for Tom. But I'm not Tom, and never was, that was just a character I played for thirty years or so. As the years go by, I've slowly started to unlearn some of the reflexive habits you develop when you immerse yourself in a role that way, so committed that you mistake acting choices for character traits. Particularly when those choices (modestly professing a distaste for the spotlight, for example) have always been warmly received by those around you.

Nonetheless, my belief in my own modesty had been getting shakier for years. Looking at myself in the mirror felt so good that it became harder and harder to believe that other people didn't get some enjoyment out of looking at me. The things I'd spent my whole life trying to avoid had all happened to me at once, and I'd survived! I'd done great! I couldn't be *that* bad, right? I still wasn't quite letting myself imagine that I could be famous and successful. But I had started to ponder the hypothesis that plenty of famous people must not have foreseen becoming famous until, one day, they were. And then, as 2021 turned into 2022, I went ahead and proved that hypothesis, by becoming famous myself.

And the instant that it happened, I realized I was ready, that I had been ready, that this was something I'd wanted all along. I first confronted the fact of my impending celebrity before anyone else knew it was coming, not even my girlfriend, Genevieve. I was down in LA, having won three games the week before, and having six days and two plane trips to think about it since then. I had done better than I thought I would, but I still wouldn't let myself believe that I could be *that* good.

But then I went to the studio that next Monday, and I won all five games I played—kind of convincingly. I didn't want to seem cocky, to myself or anyone else. But the results spoke for themselves. I was already surprisingly high on the all-time lists. And if I did the same thing the next day, which was plausible, I would be quite high indeed, among the legends, just like that. And that would mean that there would be people who would recognize my name and face for years to come, in the same way I recognized Julia Collins, James Holzhauer, Arthur Chu. I hadn't allowed myself to believe that could happen when I arrived at the studio that day, and had turned my phone off for the next ten or eleven hours. The only people who knew what had happened that day were the ones who were in the studio, and they had all signed NDAs. So after that last win, I had an hour or so to accept it for myself, apart from what anyone else would think. And that's when I realized: I was ready. The last five years of my life had prepared me for fame.

Was I incredibly lucky? Sure. But I'd been incredibly lucky before. Luck was more prevalent in the universe than I had been led to believe. Being the beneficiary of good luck doesn't necessarily mean you don't deserve it. And I deserved it. I deserved it! That was a strange feeling, believing that I deserved something, but I liked it.

So, yeah, that was the first answer I learned. How does fame feel? It feels great! A few months later, in February, I was flown out to DC, my first experience of celebrity treatment. Genevieve and I were slack-jawed the whole weekend at how amazing everything was. They flew us first class! With those sleep pods you can lie down in! They put us up in a swanky hotel, in a suite the size of our apartment, with floor-to-ceiling windows in the bedroom, and drapes that you could control with a switch from your bed! Genevieve's aunt, who lived in DC, texted her something along the lines of "Oh, well, that neighborhood isn't the *real* DC," and we just laughed. Of course it wasn't! It wasn't the real anything. It was Rich People Town, a place we never thought we'd be invited to!

But it went beyond travel and accommodations. Fame comes with a lot of prizes. Companies just *give* you stuff, right when you can finally afford to pay for it. They give it to you precisely *because* you can pay for it. It feels unfair on some level,[2] but listen: Genevieve and I both came from backgrounds where, I wouldn't call us poor, we were never in much danger of being actually out on the street. But we both knew how it felt to count the pennies in our bank account, and if people wanted to send us stuff, then by God we wanted to get it. We hired our friend Hilary to manage my Instagram, with the explicit goal of getting us free shit, and she delivered. Alcohol, clothes,[3] makeup, bath bombs, face masks. A trip to Ottawa.[4] A toaster with a touchscreen interface, where you could select the item you were toasting, and then choose from various pictures of that item to specify what level of toasted-ness you wanted it to achieve, and it played a little song when it was done. Where did the toaster come from? Why was it sent to us? We had no idea, but we immediately started making more toast than we'd ever imagined.

---

2   It's definitely unfair.

3   Although many of these were bizarrely unlike anything we'd wear. I won't name names, but one day a box arrived from a well-known shoe brand associated with skater culture. Not a brand of shoes that I'd buy for myself, but I could definitely build an outfit around them. I opened it to find . . . a sweat suit that looked like it belonged on a hungover sorority pledge at Iowa State in 2003. I'm still baffled.

4   If you go, I highly recommend getting a BeaverTail at the ByWard Market.

The only social media I'd used heavily had been Twitter; years of effort had gotten me up to three hundred or four hundred followers, which was, you know, fine, but less than I felt like I deserved.[5] But as the airdate of my first episode approached, I knew there would be some amount of scrutiny on me, so I locked down my account, and made a new, public account, with the handle @Jeopardamy.[6] The day my first episode aired, Genevieve and I had gotten an Airbnb to host a watch party. After work,[7] I headed over to the Airbnb to help set up. I was on the road at 4:00 p.m., which was when the episode started airing on much of the East Coast. My Twitter notifications started coming in. At first they were the trickle I'd been expecting. But around 4:15, I was delayed getting to the Airbnb, because, in their infinite wisdom, iOS puts their notifications in the exact same screen location where Google Maps puts the information about your next turn. And all of a sudden, I was getting Twitter notifications every second or two, in such a flood that I had no idea what Google Maps was expecting me to do. I took the next exit, pulled over to the side of the road, and turned off Twitter notifications. My first episode hadn't finished airing, and I already had more followers than I'd ever had before.

When my run of *Jeopardy!* episodes was in its second week, I was at the Safeway up on Pleasant Valley, buying our groceries for the week, and while I was checking out, two different people recognized me and told me how much they were enjoying watching

---

5   Sample tweet: "I think an underrated part of the appeal of Pride and Prejudice is it really captures the universal experience of being so fucking embarrassed by your family that you just want to die."

6   Still pretty proud of that, even if people consistently think it's @Jeopardyamy. It's a portmanteau!

7   As I called the hours in which I answered emails and Slack messages with creative explanations as to why it was going to take me a while to get back to them with the information they wanted, and then took a nap.

my run. They were maybe the fourth and fifth people to recognize me in public, something like that. I walked out the door of that Safeway grinning. I'd loved what had happened, both for the ego boost and for how nice it felt to have brightened people's days by doing nothing more than buying groceries. This was all amazing! But then I had a realization: if it ever stopped being amazing, if there ever came a time when I would prefer to buy my groceries without strangers coming up to me, wanting to talk, and potentially noticing the unreasonably high percentage of potato-based products in my cart—if that time ever came: there was no way to turn it off. This was my life now, and for some amount of time to come. There's no pause button on fame.

The week my *Jeopardy!* run ended I did a ton of interviews. I didn't have to; the *Jeopardy!* people made it clear that I was free to refuse any or all media requests. But I'd discovered my inner diva, and I wanted to put as much Amy out there as the market would bear. But I found myself saying things like "Ugh, I'm going to have to get up so early for this *Good Morning America* interview." #relatable

As the week went on, I answered the same three questions[8] in a seemingly endless parade of five-minute interviews. I had the experience that's familiar to celebrities, as well as anybody else who's worked in the service industry: saying the same phrases so often that at times I couldn't even tell whether I was saying anything at all, or just spouting a bunch of gibberish. Whether the phrase is "Thank you for your patience! If you can, please stay on the line for a thirty-second survey about the service I've provided you today," or "You know, I'd watched the show my whole life, and

---

8   How does it feel to be so successful? What are you going to do with the money? How does it feel to be so trans?

I thought I could be good, maybe win three or four games, but I could never have predicted anything like this!" it nonetheless loses all meaning.

I was speaking at an event in a hotel; I genuinely don't remember which event, or which hotel, but I know I wasn't the main attraction. I was trying to take the elevator down to the street. An elevator arrived, but it was already packed. One of the people in front of me, the one who had pressed the down button, thereby assuming the role of spokesperson for all of us in that particular lobby, waved the people in that elevator away. "We'll get the next one." I nodded, cosigning my representative's waiver of our claim to this particular elevator. The doors closed.

And then they reopened. A voice from the back, I couldn't see who:

"Are you Amy Schneider? From *Jeopardy!*?"

"Yeah, that's me!" My automated humble-gratified-pleasant response.

"I just want to say, my father passed away this year, cancer. And when he was in the hospital, we watched your whole run, and we all rooted you on together, he was such a huge fan. I mean, obviously it was such a hard time, but seeing you . . ."

The informal spokesman for the elevator occupants, holding the "Door Open" button as he'd been instructed, looked at me with the same expression as everyone else I could see: sympathy, impatience, fear of looking insensitive. The same way you would look if you were being held in a crowded elevator car so that a stranger could tell a story to a celebrity who was also a stranger to you. The same way I probably looked. I leaped on the next comma in her monologue.

"I just wanted to acknowledge that, and—"

"Thank you so much, I really appreciate hearing that. Have a good day!"

The Elevator Representative Pro Tem took his finger off the button, and gave me an appreciative nod as the doors closed.

I was invited to the White House. Not to see the president, or the vice president. Trans rights haven't come *that* far. But I'd meet with the Second Gentleman,[9] and give some brief remarks in the Briefing Room, at the same podium where Allison Janney[10] had stood in all those episodes of *The West Wing* I'd loved. The whole day was the same mixture of surreal and mundane I'd always experienced in major life events. It didn't seem real, and yet it quite clearly was.

Nobody seemed to know what was going on, all the buildings were too old to have particularly effective climate control. Doors stuck. People scrambled to handle things they had previously believed somebody else was handling.[11] The hairstyle I'd just had professionally done was torn to bits by unusually strong winds. I was a little gassy from breakfast for some reason. And yet it was also, you know, the White House. It looked just like it did on TV! And so, when I was standing at the podium and some reporters began asking me questions, I rolled with it.[12] "C. J. Cregg could handle this. And so can I." I conducted that briefing with aplomb.

At another podium in another town, I was standing on a stage in Portland, taking questions from the audience. One woman, having been called upon, began walking up to the stage. This wasn't part of the plan. I looked over at the moderator, who was looking at me with the same expression on her face as I suspected

---

9   You know—that guy, Ol' Wassisname.

10   Excuse me, I mean fellow Daytonian Allison Janney.

11   Everyone in DC always seems to be speedwalking to a meeting that may or may not actually be taking place.

12   Despite having been told a minimum of a dozen times that of *course* they wouldn't ask me any questions.

I had on mine. But it didn't look like this woman was about to stab me or anything, so I guessed we were just going to roll with it.

"Hi, Amy, I don't really have a question, I just wanted to give you these."

She held out her hands, and I knelt down at the front of the stage to take what she was giving me. At first I thought it was just some rocks, until she put them in my hands, at which point I realized that it was in fact just some rocks. They had designs painted on them, which I couldn't process in the moment. The woman explained that they were fairy rocks, which, sure, that sounded like a thing. I smiled and thanked her, of course, and to be clear, it really is the thought that counts, and I'm grateful to this person, who only wanted to give me something in return for the joy that I had given her. But all I could think was *I am* not *flying home with a bunch of rocks in my luggage*. And I didn't. I hope those fairy rocks found their forever home eventually.

When, during the course of my run airing, I was mugged—it wasn't anything to do with being famous, that's not the point here—I realized I wasn't going to be able to write some social media content I had publicly committed to. So I got on my laptop and posted something along the lines of "Just got robbed. I'm unharmed, they just took my purse and phone, but that means I won't be able to post much for a day or two."

Within the next few hours, I got a bunch of concerned messages from friends and family. "Are you okay? I saw the headline." The headline? But indeed, I Googled my name, and there they were, a bunch of blaring headlines for articles that consisted of, essentially, the tweet I had just posted. I realized that if I didn't want anything I did to be aggregated into clickbait news articles, I'd have to keep it private. My previously inconceivably massive follower count started to feel intimidating rather than gratifying.

So, how does fame feel? Complicated. That first thing I

learned—that fame feels great—is still the first answer I reach for. When people recognize me out in public and ask me for a selfie or something, occasionally they'll say something like "I'm sorry, you must get tired of all this." To which my response, always and sincerely, is "Strangers coming up and giving me compliments? How could I be tired of that?"

But, of course, I'd considered this possibility in advance, and thought of a variety of ways I could turn out to be tired of it. And many people do. Several other *Jeopardy!* champions have been quite clear that they don't enjoy it at all, and I do understand why. Whenever I'm out in public, I know that, at any moment, a stranger may come up and want to compliment me. Which again, to me, is usually a great feeling to have. But nonetheless, it is not *always* the feeling I'm in the mood for. When you're walking to the corner store to buy cheap wine at three thirty in the afternoon because you just had a sudden and drastic falling out with a close friend, you might not necessarily want to take a selfie with a stranger. But you can't show it. If you've ever had a friend that's had a negative interaction with a celebrity, you know that they just can't *wait* to tell everyone they know about it. And I felt like I couldn't afford anyone telling stories like that about me. And I mean the word "afford" literally.

Because another lesson of fame is this: if you get famous enough to quit your day job, that doesn't mean you've stopped working. It means that you are now a small-business owner, which means that separating your work life from your personal life is going to be a challenge. But it's even worse than that, because the small business you own is you.

After I quit my job, I had the horrifying realization that, despite having no experience at all in my new job,[13] I was not going

---

13   Famous Celebrity Trans Person.

to receive any training for it whatsoever. Not only would I not receive any training for it at the beginning, but as it went on, I was never going to get a single performance review. The only people who would offer critiques of my job performance would either (a) have no idea what they were talking about,[14] or (b) be people who I employed.

There was another moment, in December of 2022, a year after this had all started, that stands out for me. In the spring I had hired a personal assistant. There was still a part of my brain that was shocked by this, by the fact that German Catholic Jesus hadn't turned me into a pillar of salt for having someone help me with things I could do *perfectly well* on my own, if I just stopped coddling myself by, like, sleeping occasionally. But in any case, I'd hired her, and that December I suddenly realized I might have a responsibility I hadn't thought of. A friend of mine had worked as a personal assistant before, and one afternoon I texted her to ask whether I should be giving my assistant a holiday bonus. She said that I should, and even though it was just a text message, I could feel her eye roll, and the implied "*of course*" that she didn't type. And I realize that, while I had envisioned many possible outcomes for my life, including fame and riches, one possibility I had never planned for was that I might become an employer. But now I was.

I was my assistant's boss, and as such I had to assume that she felt about me the same way I'd felt about every boss I'd ever had. Sure, there would be a certain amount of attachment, based on how much she enjoyed the work and how she was treated. But ultimately, if it came down to it, if I didn't give her a holiday bonus, which she deserved, or otherwise failed to treat her well, she'd have every right to leave and go find a different employer who would place an appro-

---

14   Twitter trolls, mainly.

priate value on her service. As I would have done, as any employee ought to if they can. But when you own the business, you can't quit. And it feels strange to know that everyone you employ can never be as dedicated to your company's success, your own success, as you are. And they shouldn't be. You chose to start this business; you can't blame anyone else for its success or failure.

The other thing you can't do is complain. Even now, I feel reluctant to be talking about the downsides of fame.[15] I used to be a software engineer, working in the Silicon Valley tech industry. But for the most part, my social circle didn't have any other tech people in it. My friends were aspiring actors, or stand-ups, or writers; or else they were activists, or working for nonprofits. And yet, we all had jobs. We worked for an outside entity, and as such we all had certain complaints and joys and doubts and hopes in common. But when I became famous, suddenly it became so much harder to hang out and complain about my job with my friends.

The three or four people in my absolute innermost circle were still always open to hearing how I felt, but outside of that, even when I could feel their intention to hear me out, I could also always feel their unspoken[16] inability to take my problems seriously. I was a millionaire. I had become one of "those people," I had hit the jackpot, I had a level of financial security, and outside validation, that they would most likely never experience. What did I have to complain about?

And again: fair point! I wouldn't trade my life for anyone's, more or less. So how could I complain? Easy: the same way everybody can complain, all the time. The only difference is the amount of camaraderie those complaints can elicit. And even with

---

15   Again: I love it! I feel incredibly lucky!
16   And occasionally spoken.

the people in the inner circle, I'm not able to *receive* the support they offer in the same way.

I have a friend who lives in one of those giant, uninhabited states between the Rockies and the Mississippi, who has lived through more trauma than I can comprehend, with no support whatsoever from their nightmarish family, or their even worse ex, trying to keep a roof over their head with various menial jobs, and no clear path to ever breaking out of that cycle of poverty. And they are one of that inner circle who have been supportive throughout the whirlwind, who have always been willing to hear and sympathize with complaints like "Ugh, all these corporations keep sending me things for free," or "This stylist provided me with a designer dress for an awards show that I didn't think was flattering." But I suppress those complaints anyway. Even when they can sympathize, I can't receive their sympathy without guilt.

And the same applies to you, dear reader. I'm not asking for your sympathy. But I've learned this: all those famous people, who have written TV shows and movies and books about the experience of fame? They did a good job! All the clichés you see are true. It's wonderful, exhilarating, addictive. That's something I couldn't know without experiencing it, the addictive nature of fame. You see those celebrities who can't give up the fame, the Bo-Jack Horsemans of the world, people whose careers peaked thirty years prior, still going on celebrity editions of third-tier reality shows, desperate to hold on to some scrap of fame. While I hope I don't become that person, I now understand why people do it. When you're famous, people *care about your opinion*. Like, all the time. Strangers, journalists, other famous people, your agent or publicist, your social media followers. When you say something, they will all pay attention and they'll have their opinions about it one way or another. When some rando with three hundred followers, the kind I used to be, spends an hour crafting a trenchant, relevant, elegantly worded tweet on an issue of the day,[17] they are rewarded with, maybe, ten to twenty reactions. Whereas if I get drunk and type whatever bullshit is in my head, it'll be a headline on *Newsweek*. People will DM me about it. I may even get interview requests. It's extremely gratifying, and I have to work hard to remind myself that it might not last forever.

One of the most recent lessons I've learned was to let go of one of the things I kept telling myself when this all started. I never spelled it out, but it was something along the lines of "Remember, Amy, this is all fake. None of the praise you get is deserved, you just got lucky." But then in December 2022, I got a message from an old high school friend asking me to testify against Ohio House Bill 454, which would have barred doctors from providing gender-affirming health care to minors in the state.

I was in New York that day to attend the Out100 Gala, but I flew to Columbus for the day and testified on behalf of young trans people. I told the Ohio assembly members that, even though my life was going incredibly well, beyond my wildest dreams, all of it would mean nothing to me without hormone therapy. That gender-affirming medical care was literally lifesaving, and I encouraged them to make it more, not less, available.

At first I felt fraudulent being there. When I arrived, there were a ton of people there to testify, way more than they were going to be able to fit in. Many of them were parents of trans kids. And the people who had invited me there were the people doing the work on the ground in Ohio, in assembly members' offices and nonprofits. Me? I was just living in the Bay Area and being a celebrity and getting toasters sent to me, not facing many of the

---

17   For example: "Couldn't remember whether the theme song I had stuck in my head was from Full House or Family Matters. Eventually realized it was from Step By Step. So yeah, another fun night in this apartment where I live alone."

challenges they were. But when I expressed my hesitation to them they said, "Listen, you can do things that we can't because you're famous. You can bring attention to this that we couldn't get any other way. There's a lot of different work that needs to be done in any movement, and yours is valuable, too."

After testifying, I headed back to the airport and went to New York, landing around 2:00 p.m. I went to the gala that night. It was a ridiculous, swanky, drunken event with loud music and swag bags. The people I'd just met in Columbus, none of them were there. They were back in their homes, getting a night's sleep before getting up the next morning for another day of fighting against the odds to protect trans kids. I was getting my name engraved on a champagne flute while eating hors d'oeuvres. It was a strange day.

What I did to become famous was answer a bunch of trivia questions. I'm proud of how I did that, and I'm honored that people connected with me. It means so much to me that I helped out trans people whose family members hadn't seen a trans person before out in the world. And what I've come to realize is this: the testifying and the partying—they're part of the same job. It was my job to go to the party and be seen at it so people can imagine what it would be like to drink with minor celebrities.[18] I was exhausted by the end of that day. But I'm good at powering through exhaustion, especially when it's to enjoy the pleasures of fame and to do the work of using that fame for good.

So what's it like to be famous? I don't know. I'm just hoping to stay famous long enough to find out. That's @Jeopardamy on all my socials, and don't forget to check out my podcast.

---

18   Pretty fun!

CHAPTER SIX

Why Must the Show Go On?

There have been countless stories about the process of putting on a theatrical production, from *A Midsummer Night's Dream*, to *Noises Off*, to *Smash*.[1] And all of them teach the same lesson: putting on a theatrical production is a *terrible* idea, and almost certainly doomed to failure. And they're right! I've been in dozens of theatrical productions myself, and I can confirm that none of them really should ever have existed. So why do I do it?

I liked theater from the first play I ever did,[2] but I didn't fall in love with theater until a couple years later, at a Seventh-day Adventist[3] summer camp in southern Ohio. It was their annual

---

1   The star-crossed NBC show centered on the attempt to mount a Broadway production of an original musical about Marilyn Monroe, one of the most delightfully deranged things I have ever seen on television.

2   *The Music Man*, probably the best work of literature ever written about the eternal conflict between unsophisticated farmers and the grifters who want to sell them musical instruments at above-market prices.

3   Which is strange given that I was raised Roman Catholic. I never did find out what sequence of events led to me attending a summer camp that was ninety miles away from my home and run by a bunch of, as far as Mother Church was concerned, heretics.

performing arts camp, a ten-day experience where we would show up, audition for and get cast in a musical, spend a week rehearsing it, and then perform it for our parents when they came to pick us up. I knew about it because my older cousins had attended it in years past. However, when they'd been there, the musicals put on were legit, shows like *Pippin* or *The Fantasticks*, shows that people had at some point willingly paid money to see.[4] By the time I was old enough to attend, however, they had switched over to egregious, royalty-free drivel.[5] This summer, they had selected *K.I.D.S. Radio.*,[6] the plot of which consists of a bunch of children disputing the editorial direction of their school's radio station.[7]

So it was dumb, but I didn't really care, and we all had a ton of fun with it. I don't believe it occurred to any of the forty or so kids in the exclusively white cast[8] that there was any issue with the scene in which three of us performed as a mariachi band, complete with overexaggerated Mexican-ish accents. My character was a fussy Italian opera star,[9] and while I only had one scene, it included a whole solo song, which only a few of us got to do. It was before my voice changed, and I was still a halfway decent singer, and an all-the-way confident one. The song was set to the tune of "Largo al Factotum," from *The Barber of Seville*,[10] but with

---

4   Not to see them performed by children, of course, but still.

5   I have no evidence of this, but my guess is that they were never paying the royalties for the real shows and some lawyer eventually found out.

6   After what I have no doubt was an exhaustive critical review.

7   A tale as old as time, really.

8   Or, needless to say, the white, evangelical, rural adults who were responsible for the whole thing.

9   Somewhat offensive, but not, like, Speedy Gonzales levels of offensive.

10   Full transparency: I just looked up the name of that aria. But if you heard the melody you'd recognize it.

lyrics about a misbehaving dog.[11] I still remember some of the lyrics: "I have a dog and his name is-a Figaro. / He barks at the milkman, the mailman he barks—Ah, so! / Bow-wow-wow! Bow-wow-wow—Oh!" I could go on, but I will spare us both.

At this point I was twelve or thirteen years old, an age at which children have learned how to be hurtful to each other. So you might think I'd have been embarrassed to do something so ridiculous in public. But I had already learned some lessons about how to deflect bullies, and one of them was this lesson: Never play their game. Play your own. Whenever somebody tried to make fun of me I would just shrug, as if I was disappointed in them for not understanding what was really going on.[12] The key is this: you can't look like *you* think there's anything wrong. You can't buy into their framing, to admit that there's anything there for you to be embarrassed about. Never admit to being embarrassed about anything, even (especially) when you are.[13] And that's how to survive middle school: keep everyone off balance and confused long enough for you to escape.

So when I performed that song, I knew the only way to approach it was to go all in: big voice, big hand gestures, big emotions. And it was a hit; the counselors were loving it in rehearsal, and none of the other kids teased me about it at all. We had our performance for the parents at the end of the week and, not to brag,[14] but I stole the show. I got the biggest reaction, the most applause, the least obligatory-sounding laughter. It's hard to capture

---

11   You're laughing already.
12   I don't want to imply that this was completely successful. I certainly still got bullied. But it had some success, and kept me from the absolute bottom of the social pecking order.
13   When you're thirteen, you're embarrassed by *everything*. I certainly was.
14   *Definitely* to brag.

how thrilling it is to know that you've won over a live audience, particularly a skeptical one. It was my first experience of it, and it was a thrill that I'd never forget. In fact, five-ish years later, in high school English class, we had an assignment: write about a moment from your life that stands out as an extremely vivid memory, and I wrote about that performance. I'm writing about it now.

There were other things that made it such a peak experience in my life. It was a beautiful summer day, at the end of a week I'd enjoyed intensely, out in nature, with kids my own age, with songs and four square and talking about our feelings. But the main thing about that performance was that I had been good at it, undeniably. Before that day, the only thing I'd been really certain I was good at was school—homework and tests and spelling bees and such. And I was proud of all that, to an extent. But it felt too easy. I didn't feel like I deserved that much credit for academic success. But this performance was different. I'd gotten some validation from previous performances, but I'd just been doing what I was told. This time, I knew that the audience was approving of something I had created. I'd made daring character choices, committed to them, and then stood alone onstage and blew them away. For once in my life, I was actually proud.

Onstage, you can get away with anything. You can do the very things that frighten you the most, and not only can you survive them, but you can enjoy them. You can feel proud, even about the things you're most ashamed of. It's no coincidence that the LGBT demographic is wildly overrepresented among theater kids. Being queer so often means being ashamed of your queerness. It means feeling driven to express a part of yourself that is prohibited, a part you're ashamed even to have. Theater gives you a chance to express that prohibited self, right out in the open, but in a deniable way—*That wasn't me violating my prescribed gender norms! I was just playing a character! *wink**

Not only is acting a refuge from societal judgment, it can also be a refuge from yourself. I have not been a big fan of myself for most of my life. I kept a mental list of all my shortcomings, all my failures, everything I had to feel ashamed of, and I tended that list with great care, always on the lookout for opportunities to add to it. Which made it a relief simply not to be me for a while every night, not to have any responsibility for myself or my actions. Left to my own devices, it seemed like I always did the wrong thing, or said the wrong thing, and I always would, because I was fundamentally flawed somehow. But in a play, it's not up to you. The script tells you what to say, the director tells you what to do, and for a brief period you don't have to berate yourself for always making the wrong decision, because you're not making any decisions at all.

That's a lot of what theater meant to me. But when you think of actors, you probably don't think of shy, retiring, introvert types like I was back then. You probably picture big, brash personalities, always putting themselves at the center of attention. And for good reason, because those people are also drawn to the stage, and for reasons that are almost the opposite of the ones I've just described. The way I put it is that two types of people are attracted to theater: people who always want to be seen, and people who always want to be hidden. In theater, you can do both at once.

Theater is a place where you can stand alone on a stage, with hundreds of people focusing their full attention on you, and yet still be invisible. It's a place where you can say, *Hey! Everybody! Drop what you're doing and look at me! Notice what I'm doing! Stop thinking about your own life and focus on what I am feeling right now!*, and yet somehow say it selflessly, humbly, as part of a communal project. Theater brings together people who, offstage, would find each other intolerable, and offers them each what they need. For the shy, it offers escape, concealment, safety; for the confident, it offers attention, freedom, validation. In theater, not only can you do the very things you fear the most, but you can do them with the very people who make you fear it. Putting on a play is like a massive simultaneous trust fall, with everyone involved in the production constantly falling, even as they constantly catch each other. If you do the thing you're afraid of, and do it in collaboration with the people who make you afraid of it, then eventually you'll start to realize that you no longer have anything to fear.

CHAPTER SEVEN

How Did You Tell Your Friends?

Birthdays made me nervous at the best of times, and this one had more riding on it than usual. It was the first time I had ever planned my own birthday party, and it was my first birthday after Kelly left. I had the fear shared by anyone who has ever planned a party: What if nobody shows up? A decent number of people had *said* they would be there, but my mind reliably went straight to worst-case scenarios, and there were reasonable causes for concern. My birthday parties, since they were always on or near Memorial Day, usually struggled to attract guests, who often had other plans for the holiday weekend. Moreover, the one common bond among my guests was that they all lived in the Bay Area, home of free spirits, meaning that you can never assume that anybody won't flake on you until they actually walk through the door.

But there was yet another reason I was nervous in the hour or two before the party was to start, a reason that I hadn't expected when I got up that morning: this was going to be the first time any of them would see me in my girl clothes. For a few months leading up to that birthday, I had been changing from boy to girl every time I was alone. As soon as I walked through the door after work, off would come the jeans, the T-shirt with "ironic" text written on it,[1] the boxer briefs, and on would come some bright panties, a skirt, a nice tight top.

When I woke up on weekends, I would lie in bed happily thinking about what outfit I would wear that day, mentally going through the clothes I'd accumulated from brief darts into the women's clothing section at Target, clothes that I still kept in the same drawer they'd been hidden in while Kelly was living there, despite there being nobody in my apartment to hide them from anymore. That morning I'd decided on a black blouse with white polka dots and a denim skirt, along with some cute sandals. But I'd assumed that, prior to anyone's arrival, I'd take all those things off, hide them away, and put on some jeans and whichever T-shirt I pulled out of the laundry basket.[2]

But when the time came for my costume change . . . I couldn't do it. It had been clear to me since February, at the latest, that I was trans, and that as such, there was no alternative but to be open about it. If I didn't, if I kept it a secret in any way, then I would always have to keep some boy clothes around, to wear in front of whoever I was still hiding myself from. Therefore, at some point, I was going to have to let my friends see me in my real clothes. But, not to put too fine a point on it, I was fucking terrified. Or rather, *more* terrified, since just a few months earlier, every single one of my relationships had been put in jeopardy by the divorce. How many of my friends were friends with "Tom," and how many of them were just friends with "Kelly's husband"? I still didn't feel I knew the answer. Kelly had always been in charge of our social life, which had been just fine with me. I'd always considered my-

---

1   My favorite, to the extent I liked any of them, was one from H&M that just said "Not another stupid text T-shirt" over and over all the way down the front.
2   My feeling was, why put your clean laundry away in a drawer? You're just going to have to take it out again when you wear it.

self to be an introvert. Social situations gave me anxiety. But leaving Kelly in charge meant that all of "our" friends were in some sense really "her" friends.

And now I was supposed to add yet another obstacle? No longer just not "Kelly's husband," but not even "Tom"? Some new person—a woman whose name I didn't even know yet? How could I count on anyone to still care about me? How could I know that there would be anyone left to offer me support or understanding, or even just want me on their trivia team at Bobby G's on a Wednesday night? It seemed audacious to add this new burden on everyone I knew, and it was hard to believe any of my relationships would survive this new demand I was making on them.

And yet.

The people (if any!) who would come to this party were the only people left in the world who I could count on for anything, the only people I could ever turn to if things went wrong. Even if they were only coming to this party out of obligation, at the very worst I could count on that same sense of pity if I ever needed someone to drive me to the hospital, or talk me out of killing myself, or keep me off the street if I lost all my money.[3] So I thought I needed to hide this from them, lest I discover that, while supporting "Kelly's pathetic ex-husband" was doable, supporting "a weird pervert posing as a woman for inexplicable reasons" would be a bridge too far.

And so I was shocked to discover that, somehow, none of that mattered. This fear that had been ruling my life for as long as I could remember had suddenly met its match. Even if it meant that everyone who showed up at the party, the entire roster of people I

---

3   As I remain convinced, even now, that I'm going to somehow do at some point. Money is scary.

could at least vaguely consider as friends, were disgusted, repelled, confused, whatever; if everyone I knew abandoned me and I was left friendless and alone; even that, in that moment, seemed like a better option than putting on some Fruit of the Loom boxer briefs, a pair of baggy, shapeless jeans, and some fake-vintage *Star Wars* T-shirt I'd bought at Target. If that was the price I had to pay to feel any kind of human connection, it wasn't worth it. I'd been alone before, and I could do it again if I had to, but if so, I was damn well going to be alone in a cute outfit.

It was a weird position to put my friends in. I'd never given the slightest indication to any of them that I had any discomfort with my gender presentation.[4] And when they showed up and I was suddenly wearing a skirt with a cute top, I wasn't going to offer any explanation, because I still didn't have one.

I wasn't coming out that day; on some level I still didn't believe that I ever would. I assumed I would just keep living in denial of my own gender and sexuality, just as I'd been taught, and just as I believed everyone did. I had no expectations or even hopes for what their reaction would be, it was outside the scope of my imagination. But I had no choice. Their reaction was up to them.

Most of the people who said they would come came, more than I'd even hoped. My first solo birthday party was an actual party, with a bunch of cool people from different social circles, some of whom I'd never thought would show. There were no awkward silences. I'd planned a sufficient number of activities for strangers to focus on and make small talk about, while still allowing people to fall into natural conversation groups and ignore the activity of the moment.

---

4   At least, I'd tried hard not to give any indication, and at the time I thought I'd been successful.

As I realized the party was a success, it occurred to me that the only thing I was disappointed about was the very thing that I'd thought I was hoping for, which was that nobody had made any comment about my clothes. Nobody had even seemed to notice, really, although that seemed unlikely. Granted, I knew I had put my friends in an odd social situation. I hadn't made any comment on my clothes myself. I hadn't offered an explanation, or even acknowledged that I'd made a pretty dramatic change, stylistically speaking. I didn't have a coherent explanation to give, and I was so afraid of rejection that I didn't want to raise the topic myself. But it was a huge moment in my life, and when they ignored it, it left me feeling slightly disconnected from them, even as I was thrilled that the party was going smoothly.

The last guest to arrive was Jenn. She was my newest friend of the group. We'd only hung out a couple of times, and I might not have even invited her if I wasn't trying to minimize the chances of nobody showing up. I went down to the lobby of my building to let her in, we greeted each other with a hug, and she stepped back, and said: "I like your outfit!"

"Thanks," I said, blushing slightly. "The elevator's over here."

We headed up and joined the party.

It's been almost six years since that day, and though Jenn is still late to everything, I believe she is the only person from that party with whom I'm still close. On my journey through transition, she gave me so many gifts, and so much support, and I will forever be grateful to her for all of it. But the first thing she did was maybe the most important: one acknowledgment, one compliment, one sign of acceptance, was all I needed. From that moment, I knew I was doing the right thing. I knew I would be okay.

CHAPTER EIGHT

What If They Let a Girl Be in Boy Scouts?

Looking back, one of the clearest ways in which my gender manifested itself in my childhood was my intense dislike of the Boy Scouts. Although "dislike" doesn't get it across. I hated it, I feared it, I found it disgusting and disturbing. Being in the Cub Scouts I had no problem with. You just hung out in somebody's basement and did crafts. Fine. In fact, a few of the handful of actual father-son bonding moments I ever got to have came out of Cub Scouts.

My favorite thing we did in Cub Scouts was the Pinewood Derby. If you're not familiar, the way it works is this: you and your dad get a block of wood and some wheels from the Scouts Supply Store over on Shoup Mill,[1] along with other accessories of your choice. In our case, this also gave us a chance to go to the store where my dad bought miniature lead figures of Napoleonic-era soldiers and the art supplies to paint their uniforms the appropriate colors. This was one hobby of my dad's that I actually

---

1   That little stretch by the Stillwater, where Turner Road turns into Shoup Mill, before Shoup Mill turns into Needmore.

vibed with. It was creative, but it was also male-coded; my dad was part of a group of men who used hand-painted miniatures of Napoleonic soldiers and vast terrain maps divided into hexagons to play simulations of, like, the Battle of Austerlitz. Plus, it was creativity that also had rules; it was paint-by-numbers for grown-ups, and I loved paint-by-numbers. It meant that I could make pretty pictures, but without worrying that my picture was "wrong," whatever that might mean.

It was also one of the few activities that it felt like my dad did not want to share with me. I don't know why, but my memory of the worktable in the basement where he did his painting is tinged with taboo. One issue, I assume, was that I have always been a little clumsy. If that worktable were to appear before me right now, as an adult, there's a reasonable chance I wouldn't be able to resist poking around and investigating and then accidentally end up spilling yellow paint over a whole row of Austrian grenadiers. Clearly, keeping ten-year-old Amy away from that whole vicinity was only common sense. But I felt that there was something more to it. It didn't seem accidental that the worktable was set up in the least appealing corner of an already unappealing unfinished basement. Painting miniatures was something my dad did when he wanted to be alone, to forget about his family and finances and responsibilities, and lose himself in thoughts of Marshal Soult taking the Pratzen Heights. Or maybe not. It's not like I ever asked him about it.

Where was I?

Oh, right, the Pinewood Derby. So yeah, after you bought your block of wood,[2] you tried to make it into a little "race car," which would then be "raced" against the cars the other kids had made,

---

2 Pine, I have to assume.

by being placed at the top of a slanted board and allowed to roll down to the finish line. There was some basic physics to it, which was also fun. My dad didn't get to play the role of my teacher as often as my mom did, given that she was both an actual teacher and only sporadically employed, and thus had more time with us. But down in that basement we talked about how to reduce drag (in retrospect, probably a pretty negligible factor), and where to apply fishing weights to best distribute the mass of the car.[3] But we got to do some playful stuff, too. We had to decide what color to paint the car, and I'll always remember my dad telling me how British cars were always painted "racing green," and our discussion of how we would set the fishing weights in the middle so it looked as if there were people in the car.

Granted, it all got a bit tainted when the day arrived and our car didn't win. I hope I didn't make it too obvious, but I blamed my dad.[4] This felt like a class project, and I *always* won (got the best grades on) class projects. It must be my dumb dad's fault! So yeah, that's one of my favorite memories of my dad, and it ends with me resenting him for something that wasn't remotely his fault. I feel sorry for the guy. He was given the job of making a man out of me, and he was doomed from the start. But none of us knew that yet.

Which is why he insisted I join the Boy Scouts. I don't even know if he insisted, really. Boy Scouts was just the thing that came next, part of the schedule of my childhood. At a certain age you started going to middle school, and you started going to Boy Scouts, and you started being awful to your peers.

Middle school's the worst, this period between when you learn

---

3   The car had to be beneath a certain weight limit. In practice, this meant every car was exactly at the weight limit, or as near as possible.

4   I'm pretty sure it was obvious I blamed my dad.

how exciting and powerful you feel when you hurt someone else, and when you learn that you still shouldn't do that.[5] The first meeting I went to with my Boy Scout troop, some of the boys were chasing this other guy around the parking lot to beat him up for some reason.[6]

I had no idea what was going on, and I got the impression that this was just how every Boy Scout meeting wrapped up. And if there were going to be chasers and chased, I knew which side I wanted to be on. So I did my best to establish my bona fides by talking shit about him, this boy I'd never met until this moment when he was being bullied. Again: Middle school? Bad idea. I'm against it.

The main problem was that I hated boys. And the thing about Boy Scouts is: it's nothing *but* boys! It was bad enough at school; another disastrous aspect of middle school was that I wasn't allowed to hang out with girls on the playground anymore. But at least girls were *around*. Without their presence, boys were even more horrifying, myself included. They would talk about sex in the crude, hateful way that people do when they've been denied any actual knowledge of it. I remember this kid explaining that he'd had "poontang," but not "sex," and "poontang" was when you put your penis inside but then didn't move or anything. So that's what I thought for a while. It's not like there were any competing sources of information available to me.

Their go-to move in debate was to call the other person a "fag." They would compete in coming up with racial slurs for

---

5   Most people learn to stop being intentionally hurtful within a few years. Some take a lot longer. Some never learn it. Those people are often referred to as "front-runners for the GOP nomination."

6   Probably no reason.

Iraqis, among others.[7] At one jamboree,[8] I was summoned over for what I was told was an exciting event. I joined a group of boys huddled around a little ditch by the lake, where as it transpired, a boy had trapped a crawfish, and was vivisecting it, alive, with his pocket knife. That was a pretty typical scouting experience for me.

I mean, again, everything was really well-intentioned. I had nothing against our scoutmaster. He was a good person doing his best, a veteran trying to instill good character in us in a wholesome way, passing down the values he'd learned in the course of overcoming a vividly described childhood as a member of a street gang, and a rarely discussed stint in Vietnam. But he wanted to keep us boys busy, and by busy he didn't mean *thinking*.

"You should be doing something all the time! What are you doing sitting down? What are you doing with your hands in your pockets? Do something!" Where he really lost me was: "What are you doing *reading a book*?" as if he couldn't imagine anything less productive. But all I wanted to do was read. I couldn't imagine how engaging with the real world could benefit me. It didn't seem to have done me any good so far.

So as the day approached when I would be going to Boy Scout camp, an entire week in the belly of the beast, I was filled with as much dread as I've experienced in my entire life. I hated the thought of Boy Scout camp. It just felt like the last fucking straw.

---

7   This was the era of Desert Storm, a golden age for Boy Scouts. Lee Greenwood's "God Bless the USA" was the climactic event at one jamboree. People wept.

8   I always spent jamborees walking from booth to booth in such a way that anybody who saw me would hopefully think I was actually doing and enjoying all of these manly activities, and simply happened to be in transit between them right at that moment.

Birthday parties, basketball, Little League, whatever, I was *always* being sent off to be with boys, and do boy things, and I hated it.

But there was no use in hating it. It wouldn't get me anywhere. When your plan is to get what you want by following all the rules, and then it turns out that the rules say that you have to want to hang out with boys all the time and do what they do, it's kind of a conundrum! There truly was not a way out of it. Because as far as everybody else could see, Boy Scout camp was an *opportunity*. An opportunity to hang out with other boys, bond with my peers, learn valuable lessons about how to be the man I was doomed to become.

Sure, I could rebel, make a ruckus, get everybody all upset and angry at me, but I could have done that at any time. And what would the result be? Everyone would be angry at me, my parents would withdraw the scant evidence of their affection, my peers would mock me, people would think I wasn't smart, that I was a little kid, barred from the grown-ups' table. In short, I would be In Trouble, the worst possible state of being. And then they'd still make me go to fucking Boy Scout camp.

The night before I was set to leave I went to the bathroom and tried to make myself throw up. I didn't succeed. I'd read in books that you just needed to stick your finger down your throat, but it didn't work for me.[9] Eventually I admitted defeat to myself, but in desperation I went and told my parents I'd thrown up anyway. They said, "Yeah, whatever, you're still going tomorrow." They knew I was faking.

So I went. And you know, a funny thing happened: once I actually got there, accepted my fate, and began really participating . . .

---

9   Years later I would figure out how to make it work as part of a bout of buli-mia. Which then ironically gave me a really tight gag reflex, a challenge when I started giving blowjobs. But I digress.

I hated it just as much as I'd thought I would! Possibly more! It was the purest example of the way I feel about my life in general from age twelve to thirty, which is that it was like those stories you read where a cop goes undercover in a white supremacist organization or the Mafia or something for like ten years, and if they were ever discovered they'd be killed. That was life as a boy for me, except there was no police department ready to evacuate me. If it ever became obvious to the other boys how much I didn't belong, some unspecified but terrible punishment would be meted out. Boy Scout camp was that experience raised to another level. It was like Indiana Jones in *The Last Crusade* when he's retrieving his dad's diary in Berlin, surrounded by nothing but Nazis as far as the eye can see.

I don't remember much of what happened that week through the fog of existential dread and despair that suffuses those memories. In the middle of the week, our parents came for an evening and we had a ceremony of some kind. I barely said a word the whole time, although, to be fair, at this time in my life I rarely said anything to my parents. My mom would do her impression of me for her friends. *"How was your day at school?" "Good." "What did you do?" "Nothing." "Really, nothing?" "I don't know."* That sort of thing.

But that evening I was particularly mute. All I wanted was to burst into tears and say I wanted to go home, but that felt as dangerous as the undercover cop in my metaphor casually mentioning his support of interracial marriage. The next day, I got swimmer's ear, and they took me off-site to see the doctor. He diagnosed it, gave me some antibiotics and earplugs, and sent me back to camp, where I went to the one phone we could occasionally use, called my mom, and said, "I need to go home! The doctor is making me go home!" The doctor had in no way said that.[10] But I had already

---

10   I'm pretty sure he explicitly said there was no need for me to go home.

decided that this was my ticket out of there, and I wasn't going to give it up. The scoutmaster seemed disappointed, but relieved. He'd probably realized that I was not a problem he was going to be able to solve.

My mom came and picked me up, and as we were driving home she said, "I came yesterday completely prepared to take you home. I thought that that was what you would want. And you didn't say anything."

I was glad to hear my mom admit that she knew I didn't like the Scouts; there'd been a kind of grim determination among all of us to insist that, actually, I was having a great time. But I also felt betrayed. Of course I didn't say I hated Boy Scout camp! I wasn't supposed to say that! I was being good, and as a reward you drove away and left me in hell!

A week later, the proof arrived that my scoutmaster had given up on me: my merit badges for the classes I'd started at camp, but not finished, came in the mail. It was a kindness. My scoutmaster knew that I had been told by my parents that I had to stay in Boy Scouts at least until I made it to the rank of Second Class. Not a particularly challenging achievement, as the name implies, unless of course everything you're required to do makes you feel sick to your soul. My ticket to freedom was three merit badges: basketry, leatherworking, and swimming. The basketry and swimming, I did at least kind of earn. I knew how to swim and I'd woven a basket.[11] But when it came to leatherworking, I hadn't done a single thing. It was simply a gift from my scoutmaster to me, a commutation of my sentence, and I was grateful.

But you know, everything in life is a learning experience. So what did I learn from Boy Scouts? Certainly I didn't learn to be

---

11   A regular Daniel Boone, that's me!

more comfortable around men. If anything, that nightmare made it harder. I *should* have learned the value of speaking up about my needs; if I had just told my mom how I felt that night I would have been spared another day of agony, and my mom would have been spared a two-hour round trip. But I didn't.

I still haven't learned those lessons, to be honest. Men still make me uncomfortable. I try to be open-minded, and not to pre-judge people, but I prejudge men hard, all the time, even though I try to fight it. I'm trying to get better at that; I have, after all, met quite a few men who are perfectly decent people, who only barely at all remind me of that crowd of ghouls crowding around to watch an innocent crawfish being tortured to death. And if nothing else, there sure are a lot of men around, so it'd really be convenient if I was more at ease with them. I've done a bit better about vocalizing my needs, but after some fifteen years of therapy I still find myself in that same old pattern sometimes, screaming out my distress internally while sitting silently with what I hope is a pleasant smile on my face. It would be nice if Boy Scout camp had helped me socially, but every bit of guidance they offered felt like the wrong lesson at the wrong time.

No, when I think about it, looking back, there's only one les-son I learned, and even that one I didn't really fully grasp for de-cades afterward. The only thing I learned from being in the Boy Scouts is: I'm not a boy. Not even Second Class.

CHAPTER NINE

What's It Like Having ADD?

I was supposed to be writing about other stuff today. The deadline for my book is, oh God, sixteen days away. In sixty-three minutes it will be fifteen days away. And I shouldn't be writing this paragraph at all; my editor says he doesn't like it when I narrate myself writing, which I get, I totally get that. I don't like it, either! But I don't know what to tell you, I formed my intellect as a white American man[1] in the 1990s. I like my writing styles like I like my women: self-referential.[2]

I was supposed to be writing about my experience with ADD! I have many thoughtful and interesting reflections on the subject, and I had promised my publisher that I would type those reflections out for them to publish, in return for which they had given me a frankly shocking amount of money. The job description was basically indistinguishable from my daydreams: "So . . . you want me to write down all the stuff I think about myself? To actually lay

---

1   You know what I mean.
2   Sorry again! That joke doesn't even work, really. But I like it.

out the arguments I've been crafting for an imaginary audience in my head my entire life?"

But it turns out that every single story ever written is right: When your wish actually comes true, it might be great,[3] but it's also harder than just doing whatever you were doing when we first met you in the pilot episode. Better, probably! Way better. Just harder. It opens up a wider scope of opportunities for your life, and I love that, I love how many more ways I can now imagine life turning out for me. It's just that some of the outcomes becoming more imaginable are . . . not great, Bob![4] There are a lot more standard tropes I can imagine inhabiting, many of which I think I'd love. I can imagine myself as a hipster-approved queer celebrity,[5] or a panelist on *Wait Wait . . . Don't Tell Me!*, or a respected author, or a sought-after brand spokesperson. So many fantasies that I can now consider at least plausible.

But if those tropes are plausible, then *by definition, inherently,* it must also be plausible to imagine myself as other tropes as well—junkie, diva, desperate has-been, serial divorcée, complete sellout, fraud. It cannot be the case that "icon" is a plausible outcome, but "icon, but like in a cautionary tale sort of way" is not. They're the same. They are each made possible by the other's existence.

That's not what I'm supposed to be writing about! It's not even what I *intended* to write about! I opened up my laptop to write about what I'd been doing while I procrastinated on writing. See, I'm hanging out in this random Airbnb, where I went to get away from the distractions and temptations of being at home. But if one is determined to be distracted, Airbnbs offer plenty of

---

3   It definitely is in my case!

4   That's a *Mad Men* reference, a meme that was briefly viral in 2014, but has been viral in my brain ever since.

5   And, dare I say it . . . sex symbol? I dare say it, but only in a footnote.

excuses, and this one more than most. It's one of those themed Airbnbs, and like roughly 50 percent of all themed Airbnbs, this one's theme is "midcentury modern." It's got a vintage radio on the entertainment center that may or may not be functional, and it also has a record player, and a handful of assorted (and I do mean assorted) vinyl records to play on it. So instead of writing, I've been looking through these records, marveling at the art of the album covers. I grew up with them, but it's been a while, and you forget what it's like to have a stack of dozens of 12"x12"[6] physical works of art lying around your home.

I came across one album whose cover astonished me. It was a pretty standard format, just a giant headshot of an attractive young woman. Her outfit was weird, but only in the way that every outfit in the 1980s was weird.[7] What really captured my attention was the singer's expression. Her lips were parted strangely; her eyebrows were arched, but her eyes were unfocused. She looked the way people look when they freeze on a Zoom call, caught between two different facial expressions.

So of course I had to send a picture to Genevieve. And Jess. Oh, and Sam—this was just the sort of thing that Sam would appreciate. As would Ben, for that matter! So when you think about it, I really didn't have any choice but to start four separate text message threads. Which meant that there were going to be too many interruptions for me to start writing, so I played that record,

---

6   Technically 12.375"x12.375", but I feel like that slows down the flow of the sentence. Do you always hear the words in your head when you read? I do. But I also didn't want you to think I'd gotten a fact wrong, since that's kind of my whole brand.

7   Made out of some unidentifiable fabric that had probably been first invented for military uses. Having purposeless straps in odd places, as if the designer didn't understand what clothes were or how they worked. Purple. You know, '80s shit.

and then the next most intriguing one,[8] and texted my friends about them while I listened. In between texts I started investigating the singers on Wikipedia, and thought about the entertainment industry in the '70s and '80s, and learned a lot about two artists who will probably not show up on *Jeopardy!* ever, but who had really interesting[9] lives.[10] And about the way that records got me to listen to them in a way that Spotify never would have. The physicality of it does make some kind of difference.

I'm not saying albums sound "better," but then, what does it mean to say that one recording sounds better than another? In the digital world this isn't as obvious, but the reality is that all recordings, of any kind, are fundamentally *different* from what they are recordings *of*. It's just easier to see that in analog recordings because each physical copy of the recording is different from every other copy. Maybe one copy is scratched, or one is missing its protective sleeve, or one got wine spilled on it.[11]

But my point is, it's not *just* the physicality of the world that creates that fallibility, that renders recordings different from events. Digitized information is nonetheless information, and one aspect of information is that every communication has a mis-

---

8   *The Look of Love*, Claudine Longet.

9   That first album was *Hollywood, Tennessee* (what a title!) by Crystal Gayle. And get this, it turns out she was Loretta Lynn's kid sister! I'd never even heard of her, but she had a song that reached number two on the *Billboard* Hot 100. Not on this album, though. It had two Bill Withers covers, and I honestly cannot even tell how I feel about them.

10   And that's nothing compared to Claudine Longet! She was married to Andy Williams for fourteen years! Then two years later she was convicted of negligent homicide after shooting her new boyfriend! Who was an Olympic skier! *Saturday Night Live* publicly apologized to her for a joke they made about her in their first season! How did I never hear about this?

11   But not any of the ones in that Airbnb. Not while I was there. I swear.

take in it. Everyone mishandles situations all the time, everyone sometimes manages to fail. Either they fail to communicate the ideas that they are having,[12] or they fail to comprehend the ideas that other people are attempting to communicate to them. Even without human error, mistakes always happen. Cameras routinely capture images that don't resemble reality at all.[13]

And I guess one benefit of this particular form of analog media is that it forces you out of that digitized way of thinking. If I listen to a record, I am listening to a record. If I'm listening to Spotify on my headphones, then that's just . . . a thing that is happening, like I've changed the lighting in the room or something. And that's fine, I like listening to Spotify; I do it all the time.[14] But it's not something I can really focus on. A record somehow demands your focus. You find yourself responsible for it. And not just in the sense that you're the one that put it on, but also that you have taken temporary custody of a relatively delicate physical object. If some dust or something gets stuck on the needle, you need to hear that and intervene to prevent anything from getting damaged.[15]

Of course, this caution not being necessary for streamed music is a *good* thing. It is so much better in so many ways.[16] But just like my dreams of stardom, the amazing miracle that is streaming music also has some unavoidable downsides. And one of them

---

12   I may be doing this right now.

13   As everyone knows who's ever tried to sext.

14   I'm doing it now, as I edit the original draft of this chapter a couple weeks later.

15   Right? Something like that could happen, right? I don't know, I grew up with records, we didn't get our first CD until I was maybe twelve. But that means the vast majority of the time I've spent playing records has been as a child, a child who'd been given very clear instructions about the delicacy of the equipment involved, and the necessity of using it properly.

16   Spotify tracks are so rarely damaged by dust, for example.

is a certain disconnect. You are not responsible for the music; if anything goes wrong with it, it has nothing to do with the music itself. With digital music, it will never be the case that you'll say, "Oh, sorry, I can't play 'Red (Taylor's Version)'; my copy got all messed up somehow." You might say, "Oh, sorry, playing 'Red (Taylor's Version)' is one of many things I can't do, because my phone died."[17]

So anyway, all of that is an illustration of what it is to have an ADD brain.[18] I want to explain it rather than illustrate it, but I can't make myself do it. Because you know who's bored of reading about the experience of living with ADD? People with ADD! That's the exact sort of discussion that would bore us. Which makes it a challenging topic to discuss.

People with ADD are fascinated by ADD, until they're not. And so, yes, I went through a time where I watched endless ADD explainers on YouTube, and all of that.[19] But ADD hates being told something it already knows.[20] It doesn't just feel irrelevant; it feels painful.[21] But on the flip side, being told something you

---

17   I mean, you wouldn't say that. You would only say something like that if the subject of "Red (Taylor's Version)" had been brought up, and if it had you would refer to it by saying, "I can't play that," or "I can't play that song," or whatever was contextually appropriate, not repeat the title.

18   Oh damn! I'm writing about it after all! Now I can claim I was writing about it the whole time! Or maybe this isn't really illustrating my ADD, but some other personality trait of mine that I'm just lumping in with it, just to save myself from having to write something else about ADD like I said I would.

19   I really loved "How to ADHD."

20   As you would know if you'd watched those explainer videos.

21   Although here my perspective is admittedly skewed, since I've made "person who knows a lot of things already" such a crucial part of my identity.

don't already know? That feels amazing! Learning was my first drug, my oldest form of self-medicating.

Most of my life I never considered that I might have ADD. I thought of ADD as essentially meaning "ill-behaved," and I certainly wasn't ill-behaved. I behaved great. I crushed it at being well-mannered. It was a lot like being trans. I never thought it could be true of me, because I was a good kid, and kids with ADD, or trans kids,[22] those were bad kids. Having ADD wasn't quite as bad as being trans, to be sure.[23] But most people thought they were both just euphemisms for being perverted or insolent. Certainly my mom was vocal in her belief in the nonexistence of ADD.[24] Some kids were just naughty, was how she looked at it. Though to her credit, she never judged them for it. It wasn't kids' *fault* that they were naughty, and she understood why people had felt the need to invent a term for it that sounded less judgmental. She just wasn't going to pretend it was something different. And as a teacher, she saw accommodations for ADD as just another excuse, an attempt to get an advantage over the other students.

Most people[25] agreed with her back then. By the time my brother got diagnosed with ADD in college, my mom was ready to understand the point of the accommodations. She had seen throughout his academic career that his grades consistently underrepresented his understanding of the material. He just had to consider everything from every angle. When he would take a test, he would get a reasonably high percentage of correct answers . . .

22　Or "faggots," as they were referred to back then.

23　Having ADD didn't require forcing adults into talking about, you know, *ess-ee-ex.*

24　My dad usually handled the disparagement of trans people.

25　At least, most people who were parishioners of Corpus Christi Parish in the 1980s.

on the first third of the test, which is as far as he would get before time was up. Giving him longer to finish the test wasn't about giving him an advantage over the other students, it was giving the teacher the opportunity to better assess what he had learned. My mom's reaction surprised me, although I had long felt that ADD was a perfectly legitimate diagnosis. I mean, for other people, not for me. I knew my issue: I was just plain lazy.

As I saw it, my story went this way: I happened to be born with a brain suited for school. All through grade school, I had the best grades in the class, and I didn't even have to work for them. Then when I got to high school, the work grew a little more difficult. And I realized that I could either start working hard, and continue having the very best grades, or continue half-assing things and just have very good grades. And half-assing it seemed like the way to go, so that's what I did.[26] But that wasn't a convincing story.[27] I remember sitting in my living room my freshman year, looking at a geometry assignment and crying. I was crying not because the assignment was hard, but because it was easy. Math came to me particularly easily.[28] There was no reason why I couldn't just do this assignment. And yet, I wasn't doing it, and I couldn't understand why. Laziness was the only reason I could come up with.[29]

Having undiagnosed ADD *sucks*, y'all. What I've learned about it is that if you have ADD your brain doesn't process dopamine very well, so the kind of default background levels of dopamine

---

26 I remember talking to my friend in Spanish class about some assignment, and she was like, "*Tomas, ¿Por qué nunca haces tu tarea?*"

27 Much like my "I'm not trans, I'm just a boy who wishes with every fiber of his being that he was a girl" story.

28 Up until I took Differential Equations in college, at which point it just completely stopped making any sense.

29 At least, the only one that placed the blame where blame should always be placed: on me.

we get in daily life just don't do the trick. You know how you feel when you're really, really bored? That's my default. In Dungeons and Dragons terms, I always take +2 damage from boredom relative to an unmodified character. That's why it's treated with stimulants; the stimulants provide your brain enough dopamine that it can feel the same way a non-ADD brain feels all the time. Stimulants like Adderall[30] don't make me feel wired up and energetic, at least in moderate amounts.[31] It makes me feel calm, stops the raging of my brain's constant demands for more stimulation.

But while it doesn't make me feel wired up, it does make me feel like I have a superpower. The first time I tried it, I was in my messy apartment and noticed some underwear lying on the floor, and thought, *I should put that in the hamper.* And, get this, *I actually did it.* I just picked it up, carried it over to the hamper, and put it in, just because I'd decided to. I was blown away. An entirely new experience for me. Is this how other people lived their lives? They could just decide to do something, and then do it, without having to argue with themselves for a while?[32] I saw the answer to a mystery that had always baffled me: How did people keep their homes clean? I'd never been able to imagine what that would be like.

That's how I got diagnosed. Genevieve has ADD, and we hadn't known each other a week before she said, "I think you have ADD, too," basically just by having seen the state of my apartment.[33] And it's not just that ADD makes cleaning harder, it also makes it less rewarding. Having ADD means that you never fully

---

30 Or, say, cocaine.

31 Which is harder to get right with cocaine, I have to admit.

32 Potentially for years.

33 I.e., "looking like someone had broken in and ransacked it in search of a microchip with the identities of every organized crime leader in Gotham."

develop object permanence.[34] When you can't see something, you'll lose all knowledge of what it is or where.[35] So having piles on a shelf might look messier than having everything tucked away in a drawer, but if something is tucked in a drawer I might as well not have it.

And that wasn't the only evidence Genevieve had seen. Part of the mess in any ADD house is abandoned projects. I've got a bunch of painting materials somewhere, a partially finished adult coloring book somewhere else, perhaps under my half-completed latch-hook wall hanging. I just came across my guitar and thought, *Boy, I haven't picked this up in forever.* Then I spent half an hour trying to tune it, replaced one of the strings, then put it down and haven't picked it up since. I've written the first third of multiple screenplays. I made a stop-motion animation film once that's five seconds long. I get into something, buy all the supplies, do it for a bit, and then hit a wall with it and don't care about it at all. For that matter, I abandoned this whole essay a thousand words ago.[36] I'm only finishing it now because my editor is making me.[37]

And it interacts with my depression, too. When it comes to mental health,[38] everything is connected. As I've been saying, my ADD gave my depression so many great talking points when it wanted to recite a list of my failings to me while I was trying

---

34   I learned this on TikTok, and the example they used was how they never put anything in the crisper drawer of their fridge, because if they do it will simply stay there forever and rot. I'd always been the same! I called it the "Drawer of Oblivion."

35   Last time I moved I found *so many* drugs. I'd hide them because someone was coming over, and then forget about them completely. And this is drugs we're talking about!

36   I'm not in that Airbnb anymore.

37   And, again, paying me just an absurd amount of money to do so.

38   Or literally anything else.

to sleep. And depression in turn takes away your interest in the world around you, which means you don't have the dopamine your brain needs, and you end up turning to the easiest sources you can find: masturbation, junk food, drugs, binge-watching *Below Deck Mediterranean.*[39] And then all of those become new failings to add to the list! And while meds help mitigate the worst of it, they don't really fix anything.

Then again, how much of it is something that needs to be fixed? I mean, there's no way I would have been so good at *Jeopardy!* without ADD. ADD made me literally addicted to learning. My brain needs stimulation and that's one of the best non-self-destructive ways to get it. I'm not great at learning a lot about any one thing, because that gets boring, and I end up tossing it on the pile with my juggling equipment and Rosetta Stone CD-ROMs. But *Jeopardy!* doesn't reward that level of detail anyway; it values variety.

I value it, too! I *like* all my abandoned projects. They each gave me what I wanted out of them. I wrote like six songs in college and haven't written one since, but I'm still proud of them. I know that it's possible for a language to have a way of indicating number that isn't just "singular" or "plural."[40] If someone were to say to me, "Okay, we've got this new treatment that completely eliminates whatever this ADD thing is," I wouldn't take it. I like having the Adderall that fixes the most inconvenient part of it, but fundamentally, I don't want it to change. I'd still rather have it

---

39 Just for example. I watched one season of it because the promos during the preceding season of *Top Chef* heavily implied that two of the women on the crew were gay. And they were not at *all* gay, but by that point I had to see how Hannah's fight with Bugsy turned out.

40 Finnish has a separate form just for things that are in pairs, like legs or sides of a coin or romantic partners. Well, those don't necessarily come in pairs in the Bay Area, but you know what I mean.

untreated than have it taken away. So is that a disorder?[41] When you're talking about your internal mental processes, what's the difference between a disorder and a soul?

Bonus Extra-Long Digression

You may be wondering why in this piece I keep saying "ADD," even though it is properly called "ADHD." Well, I have a lot of thoughts about that, and they tie into the whole question of what is "ADD" and what is "just how I am." So here's the history of that terminology change.[42]

So, for a long time, people have noticed that there was a certain "type" of person that existed often enough to be noticeable, who were characterized by traits such as being easily distracted, restless, forgetful, or hyperactive. Of course, all of these were traits that anyone might have, but still, it seemed like there was a tendency for them to cluster together in the same person.

Eventually, modern psychiatry came along and decided that some personality "types" could be considered "disorders," if they made it harder for that type of person to function in modern, Western society.[43] They decided to call this particular type "attention deficit disorder," indicating that what made these people "disordered" was that they had difficulty paying attention to things they didn't find interesting.

---

41   I mean, it has to be, because if I was taking Adderall just to make my life better and not to "treat a disorder," then it would be evil drugs and not good medicine. And it's already criminally hard to get Adderall even in its "good medicine" persona.

42   Or at least, my understanding of that history. I fact-checked none of this! Do not cite this in your homework assignment!

43   The thought of making changes to society in order to allow these people to function in it did not, in general, occur to them.

But then psychiatrists changed their mind about what should define this type of person. They decided that people who had trouble paying attention weren't really disordered, as long as they could just sit still and be quiet. So when I was a kid, that's what the general public vaguely knew: there was a thing called "ADD," but what it meant was that you were rowdy and got in trouble a lot at school.[44]

In 1987, psychiatrists officially changed the name to "ADHD," reflecting this new understanding, and that is the official name to this day. Of course, most people don't pay close attention to the psychiatric literature. But, over the ensuing decades, more and more people have heard "Oh, ADD is ADHD now. The *H* means 'hyperactive,'" and internalized the message that hyperactivity was an essential part of the diagnosis. Which is exactly what that name change was meant to accomplish.

But! Before long, psychiatrists decided that they'd been right in the first place, and that people who couldn't pay attention, regardless of whether they dealt with it by sitting still and daydreaming, or by moving around and distracting the other kids in class, were still the same type of person, with the same "disorder." And in 1994, they officially declared that the disorder now had subtypes, one of which did not include hyperactivity. And yet, bafflingly, they didn't change the name! And so most people who have learned of the name change, with its implied[45] emphasis on

---

44   And also that you were a boy, probably. As always, gender is part of the story; girls who are hyperactive are often punished and shamed for behavior that, in a boy, is considered normal (girls are "disobedient," boys are "rambunctious"), and the way they express their inattentiveness is naturally shaped by those pressures. Thus, by focusing on hyperactivity, psychiatrists were also focusing on boys, who became overwhelmingly more likely to receive a diagnosis, and therefore treatment.

45   And sexist.

hyperactivity, learned of it *after* psychiatrists had decided that people with the disorder weren't necessarily hyperactive at all!

So that's one of the reasons I used ADD in this piece: I think
it more accurately reflects the fact that some inattentive people[46]
are, for example, perfectly capable of sitting still in class, and not
getting in trouble, despite being bored out of our fucking minds.
But honestly, that's not even the main reason. I don't even know
why I told you all that.[47] The main reason I prefer the term ADD
is that it is so much better as a word. When said aloud, the *H* basically turns it from one word into two. "Adeedee" is a lovely little
dactyl,[48] rolls trippingly off the tongue. But now we're supposed
to say "Ady Aitchdee?" That's so boring! I don't want to do it! I
don't have time to say a whole 'nother word! In short, I prefer the
term ADD because I have ADD. It's only a minor inconvenience,
but it is one, and ironically it is most inconvenient for the very
people it describes.

Well, anyway, if you cared, that's why I did it. And if you didn't
care, at least it's over now.

---

46   Me, for one.
47   Because it was fun and I wanted to.
48   Or maybe a bacchius?

CHAPTER TEN

What Is Polyamory?

I met Kelly in 2004, and it was quickly apparent to both of us that we were going to get married. Around Christmas, Kelly said, "Don't propose until after I graduate in May!" If she hadn't, I'd probably have proposed by New Year's Eve, and she would have said yes, because we knew we were going to spend the rest of our lives together. We also knew already in those first few months together that we didn't want to be monogamous.

For one, we still had so much of our lives to live! I was twenty-five and she turned twenty-two twelve days after we met.[1] And while she'd had a bit of sexual experience, I had none whatsoever.[2] I'd never even been kissed. She was my first, for basically everything, and while I had no doubt that I wanted to marry her, I

---

1 Twelve days into a relationship was early to have to buy a present, but I thought I did a good job. I got her a first-edition book of poems by Sara Teasdale, an obscure poet who I'd somehow found out about and really liked. She thought it was lame, and never read it. In future years she would say, "Remember that terrible gift that you got me?" Disconnection like that was a much bigger problem than having sex with other people.

2 Well, basically, but see chapter 18.

couldn't help but realize that marrying her was supposed to mean that she would be my last as well. I was supposed to promise that I would go to my grave without ever knowing what sex was like with anyone else but her. And that made me uncomfortable. Not miserable, nothing that wasn't dwarfed by how happy I was to be with her, just a little uncomfortable.

One of the things that brought us together was our shared trauma from being raised in a culture that didn't just condemn sexuality, but essentially denied that sexual pleasure existed. With each other, we'd finally found someone to whom we could safely admit that, actually, we kind of *liked* having orgasms, and all the various activities that tended to produce them. It was such a relief to be able to say that! To be able to talk about the fact that sometimes we got horny, that we masturbated and fantasized and lusted and envied, that what we fantasized about was not the Sacrament of Holy Matrimony, and that our orgasms didn't seem to have anything whatsoever to do with Christ's love. With each other, we could talk about our *actual* fantasies, whether or not they were sanctioned by the Council of Trent or whatever.

Another of the primary ways we connected was through our fascination with pop culture, so inevitably, we got around to sharing which celebrities we fantasized about. And we quickly realized something: if either of us ever got the chance to act on our fantasies, if Kelly had an actual opportunity to have sex with Jude Law, or I had an actual opportunity to have sex with Justine Henin-Hardenne,[3] then not only would the other spouse be okay with it, they would be genuinely happy! We loved each other! If

---

3   Don't judge, she was badass. She was just as thin and white and blond as Sharapova, or more relevantly, Kournikova, and she just didn't care. She was there to win at tennis and that was it.

our partner was able to fulfill one of their dreams, why would we do anything except cheer them on?

And then of course not all of our crushes were celebrities, and we weren't going to lie about that, either. Obviously we had crushes on people we knew in real life, people who might return our interest in intimate physical interaction, unlike Justine Henin-Hardenne.[4] So why should we try and prevent that from happening? Everybody says it would be bad, but everybody also said it would be bad to have premarital sex, and premarital sex was fucking amazing.[5] So . . . why not just let each other have sex with other people? Even people who weren't in *The Talented Mr. Ripley*? And when I asked myself that question, I couldn't find any downside. It just seemed like another aspect of the question to which nobody ever seemed to have a satisfying answer: *Why is there this special category for behavior involving genitals?*

Having sex be the dividing line between okay and not okay seemed so random. Why should it be unbearable if Kelly got drunk and hooked up with someone attractive at a party? Whereas, if she was spending all of her time with that other person, prioritizing them over me, keeping them secret from me, but never had sex with them, that's not cheating? If she creates a profile on OkCupid[6] and puts "Looking for my next husband" in her bio, that's supposed to bother me less than if she went on a work trip and hooked up with a coworker one night? Doing that wouldn't mean she wanted to leave me. I knew that for a fact, because I could imagine myself

---

4 She was too busy preparing for Wimbledon.

5 Literally. <Rimshot>

6 At the time. Feel free to substitute Tinder, Feeld, Seeking, whatever's popular when you read this.

having sex with all kinds of other people,[7] and in none of those fantasies did I want my relationship with Kelly to end.

So, since we didn't see a point in being monogamous, we decided not to be. In the Bay Area in 2023, that's not hugely shocking, but in Cincinnati in 2006, people thought it was bizarre. And not just the usual suspects, the people who considered every behavior bizarre unless it would have been considered acceptable in a Dominican monastery eight hundred years ago. We didn't tell many people about it, but we didn't hide it, either,[8] and even our friends were skeptical. Which was baffling to us, but then, we were used to being lonely geniuses, misunderstood by those who were beneath us.[9]

So, Kelly made a pass at a guy that she was doing a play with, he accepted, and they hooked up for a little bit. And it was fine. Neither of us had a problem with it.[10] We were always in agreement that as long as we were communicating about everything, as long as we weren't keeping secrets from each other, it was all good. And it was! And then we moved out to the Bay Area and found a whole community of people who had reached the same conclusions as we had, who also found the concept of monogamy baffling. Our running joke was that we'd be watching movies or TV shows and say, "This whole thing could be solved if they were just poly!" Every love triangle suddenly isn't a problem, it's an opportunity. Why choose? Why not be with both of them?

---

7   Most of them, really.

8   How else would we have found people to fuck?

9   Apart from our shared trauma, we bonded over how much we enjoyed being smug and looking down on other people. *Other people are so basic and they experience jealousy, ha, ha, ha! What simpletons.*

10   Though the guy did, I think. He really seemed to be confused by my disinterest in, like, challenging him to a duel.

And sure, all the things that people warned me would happen, did happen. Kelly slept with way more people than I did.[11] The circumstances that had kept me a virgin until twenty-five were still there. I was still uncomfortable with myself, and just generally bad at dating. First dates will never be my strength, I don't think. When it comes to other people, I don't have a setting between "at least slightly uncomfortable" and "almost disturbingly comfortable," between "coworker" and "Elena Ferrante character," and neither of those settings are ideal for a first date. But Kelly didn't have that problem, she could talk to anyone. And to make our situation more uneven, I wanted to date women, while she wanted to date men, and dating men is easy, if you're not too picky. It was like she was bowling on one of those lanes they set up for kids, with cushions in the gutter. She was bound to hit *something* if she tried.

For a while after we moved to Oakland, she was sleeping with this guy named Cal, although Cal may not have been his real name. He was a pretty shady dude, had been in jail, and gave conflicting answers about his personal history. But he was also a fascinating person, quite thoughtful in his way. He went out of his way to get me to like him.[12] Cal was the only reason I managed to take advantage of the permission Kelly had given me. There was this woman Arianna, who Cal had known, and who had previously had a threesome with him and Kelly. She was cool. We all hung out and smoked weed and stuff. One time she was over, and when she went into another room, Cal said, "You should go in there and fuck her."

---

11   Not to imply that it was some huge number, but it was definitely higher than mine.

12   To a certain extent. He had some of the same "Why aren't we fighting a duel?" vibe that all the guys Kelly slept with had, but he was content to just roll with it.

I laughed nervously. "What?"

"Yeah, dude. She thinks you're cute, she likes you, she wants you to make a move. She's hoping you'll go in there and try to hook up."

And because I trusted him,[13] I did. I made a move, she reciprocated, and we had sex. And that was something that had never happened to me in my life. Kelly was the only person I'd had sex with, and she had made the first move.[14] That afternoon with Arianna permanently helped my self-esteem. It's still the case that, whenever I think that someone might be interested in me, my first instinct is to assume that it's not true, that I'm just seeing what I want to see; that I want them to be attracted to me, so I'm just making up a story that they are. But now, thanks to Cal, I can sometimes riposte: *Maybe, but I wasn't making it up that time with Arianna.*[15] He pushed me to take the leap, to initiate something, and the other person[16] had been receptive. So, that was good.

Not that everything about my wife's sexual relationship with a mysterious homeless felon that she met at a comedy open mic was perfect. Poly relationships require good communication and discussions around boundaries. And one of the boundaries we'd agreed on was that Kelly's partners would use condoms; I usually did as well, but depending on where she was in her cycle,[17] we

---

13   I mean, within parameters. He wasn't generally reliable, but in the field of "getting women to want to have sex with you" he was pretty impressive. I saw him go from meeting a woman for the first time to sucking her nipples on a public street within the course of one cigarette.

14   It took a roughly eight-hour-long date for me to finally figure out she was making a move, but she stuck with it.

15   Boom chicka wow wow.

16   She was really hot, by the way. I would have thought she was way out of my league.

17   And also, of course, depending on how drunk we were.

might decide to chance it. But we'd agreed that her other partners would always use them, or at least make a good faith effort to pull out. And yet one day I went to throw away an empty bag of Doritos or something, and when I lifted the lid of our garbage can, I saw an empty box of Plan B, nestled there on top of a bunch of junk mail and candy wrappers.

So yeah, that made me angry. Not only had she apparently broken her promise to me, but she had done a shoddy job of covering it up.[18] At least, I think it made me angry. Anger doesn't come to me easily, and I remember actually being kind of happy, glad that finally there was something that I could be certain I had the right to get angry about. For once, I could be mad at Kelly in a way that wouldn't somehow ultimately result in my apologizing to her for being mad. For once, I could expect an apology. And I got it, too. Although then she kept letting other guys cum inside her.

Eventually the relationship between her and Cal deteriorated, and they broke up. Or rather, I broke up with Cal on her behalf. She insisted. I didn't want to do it. I liked Cal! Plus, I hoped he'd get me laid again.[19] But nonetheless I found myself standing out on a corner by a 7-Eleven, explaining to Cal how Kelly just needed something different . . .

None of this made me feel any desire for monogamy, and both of us continued to pursue other relationships from time to time,

---

18   Believe me, if our roles had been reversed, that box would have been way down among a bunch of coffee grounds and chicken bones. Or more probably, in a public trash can, and not one near our house.

19   Later she said to me, "Yeah, Cal wanted to bring you to a party where there were going to be a bunch of Berkeley girls all hanging out naked. Of course I told him you wouldn't want to go to something like that." I'm still awestruck by how wrong she was. She might as well have said, "He wanted to give Cookie Monster a whole bunch of cookies, but of course I told him that Cookie Monster wouldn't like that." Good Lord.

although it still wasn't always great for me. One time, she was out at a New Year's party with some other stand-ups, texted me around 12:30 to say she was coming home, and then I didn't hear from her until she walked in the door around 4:00 or 5:00 a.m. I genuinely thought she was dead. I mean, I had a tendency to think that anyway given the slightest excuse, but this was another level. I actually got to the point of: *Fuck, I guess I have to be the one to call her parents. How am I going to break it to them? What do you say . . .*

But she was fine, of course. She'd just been having sex with this comedian she liked. The usual roller-coaster ride ensued, they had sex a bunch, we invited him on our podcast, she had a pregnancy scare with him, they broke up, she told me what a shitty person he was, we ran into each other at comedy events, and he made fun of her weight. The usual.

Then in 2016, around the beginning of the year, she started seeing him again. And on the Fourth of July,[20] she said she was leaving me. She was in love with this guy,[21] and needed to be with him, and he demanded monogamy. And she was fine with that. In fact, she wanted to be monogamous, and now she would. But with him.

I promise you: I was angry this time. Hurt, betrayed, despairing, furious, disbelieving, shocked, disoriented, bitter, everything. It was just about the worst I have felt in my entire life. And it was what everyone had predicted for our "open marriage" back in Ohio, over a decade before, what men had continued to predict to me ever since: you can't have an open marriage, because your girl will leave you for another guy. This was what they'd been trying to warn me about, the fatal flaw with my plan. And yet . . . it didn't

---

20   The same day Kevin Durant announced he would sign with the Warriors.
21   Who was living with his parents in Miami at this point, btw.

change my mind. Not even a little bit, even in the worst moments. I still couldn't see the point of monogamy.

I still can't. I still think that being poly is the way to go. Yes, Kelly left me for a guy she was sleeping with,[22] but people in monogamous relationships get dumped for other people all the damn time! This thought isn't original to me, but: everybody's relationship is polyamorous. Some of them just haven't found out yet.

And yes, that's overly cynical, but there's some truth to it. Ultimately, if my wife would rather be with someone else, then she'd rather be with someone else, and I don't see how monogamy could change that. I don't want her to be shackled into marriage with me. Maybe that comes from witnessing my parents' marriage. Divorce was not even an imaginable option for them, like most people I grew up with. I didn't know any kids whose parents had gotten divorced— that only happened to bad, non-Christian people. My dad, in particular, was always going to follow the rules, and not getting divorced was one of the biggest rules of all. And he was miserable.

My mom and dad did not appear to like each other. For most of my life, there was rarely a week that my mom didn't sleep on the couch a night or two.[23] Signs of affection were sporadic, and often felt forced. My dad was mad at my mom all the time—not angry or yelling, but clearly resentful of her. I often wished they would get divorced. I still wish they had. I know both of them felt that part of the reason they stayed together was for me. But they were wrong. Getting divorced would have been so much better for me, not just them.

---

22 To be clear: Kelly broke up with him a few months later. I was right in thinking he would be a bad boyfriend for her, but I have to admit that Kelly was right on the much more important point, which was that our marriage needed to end.

23 Or seven.

I could be wrong, but I have no reason to think that my dad had sex, or any physical connection with a woman, in the last thirty years of his life. And that sucks. We weren't close, but I always wished him well in a distant kind of way. He was trying hard to do the right thing all the time, and he deserved better.

And look, I can only speak for myself. This isn't my policy prescription for America. But it is true for me that I'm happiest being polyamorous. I am so grateful to know what different types of breasts feel like and to see the variety of labia and vaginas out there, how they react to stimuli, to be exposed to a broader range of the human experience. I'm fascinated with the female body, albeit in a complicated way.[24] I want to become an expert on it, and I intend to keep trying!

While I haven't gotten that much better at knowing when to make a move, I've at least been able to practice. And the few times I've succeeded, it's changed my outlook on myself for months at a time. I felt incredible afterward, more capable of doing whatever it is that I'm trying to do, not just sexually. It helps with the hatred I still have of my body and myself as a physical and sexual object. It's way better than it was before I transitioned. But those feelings don't go away so easily. And being able to make that physical connection with people helps.[25]

---

24 *Really* complicated. You could write a book about it. This book, for instance.
25 Which I feel bad about claiming as something positive. Because of how I was raised, wanting to feel better about my body feels wrong, shallow, unimportant. But it's important! One of the many things I've learned in transition is how awful it is not to like your body. If you think your body is disgusting and repulsive, then it doesn't just make dating harder, it makes everything harder. In our society, anyone with even a tangential connection to femininity has a lot of shame thrown at them. When I transitioned, the first rush of freedom was exhilarating. Finally, I thought, I could express my femininity, learn to love myself as a woman, start loving my makeup and hair and clothes, and . . . Wow, I sure

You simply can't get the same validation from a long-term partner. You've heard it all before from them. It's never a surprise. With other people, it's a surprise! When they say yes, it means something different. When your long-term partner says they like your body, they might just mean that they love you. Which, sure, that's nice, but you knew that already. But when you hook up with somebody who doesn't know you well, then it must be because they're horny for you, and that's gratifying on a whole different level. Especially as you get older. Once your body starts sagging, it's all the more important to have that experience of acceptance and to give it to other people. When you find yourself being attracted to people your age, people with the same flaws that you think are so terrible in yourself, you realize that maybe you're okay, too.

So, I don't know. I do in fact believe that everybody should be poly, but I know that most people don't want that, and that's their decision to make. They have their reasons. I've learned that some of those reasons are more valid than I expected. There's a difference between emotional and sexual monogamy. Emotional polyamory is harder. I know now that there were moments when I didn't feel okay the way that I was telling myself I did. I did feel jealousy with Kelly from time to time, but I just filed it in my already overflowing "To Be Suppressed" folder. I'm not immune to jealousy or above it. Jealousy is real, not just a weakness. Someone like me who has a tendency to tell people what they want to hear, and then talk themselves into believing it, can fall into some traps.

But despite all of that, all of those caveats, all of those down-

feel fat all of a sudden! So that's a whole new fight to have with myself, especially transitioning in early middle age. I never had a female teenage body. I know, I know, a female teenage body comes with its own set of challenges. But it would have been nice, you know?

sides, I still come back to this: sex isn't magic. It's simply one of the ways that people interact with each other. To me, the poly thing is part of a broader conversation about the weird way we talk about sex or don't talk about it, the power that we give it in our society, the significance it doesn't deserve. It's not that big a deal! But the idea that it is was a crucial part of keeping me closeted, keeping me isolated from myself, and thus from everyone else, and I refuse to cut myself off ever again. Maybe I'm just selfish, maybe I'm a bad girlfriend, afraid of commitment. But I feel like I've lost so much time.

CHAPTER ELEVEN

What Is the Greatest Animated Television Show of All Time?

I grew up with a strong desire for invisibility. In large part, this was due to the ever-present feeling that I was failing at performing my gender. The whole boy thing was just so exhausting, and I never felt like I got it quite right. I was always on the verge of being exposed as unmanly, and I had no idea how to avoid it.[1] Failing at boy-ness was an unforgivable sin, so my only hope was to not be noticed. Each new encounter with another human being could be the one where I slip up and have my cover blown, and be punished, possibly with violence.[2] And so I began limiting those encounters whenever possible.

---

1   That I was *not* a boy was not a possibility I considered; I assumed that other boys felt the same, found it just as exhausting and unbearable. They had just somehow managed to put on a convincing show. You'd think they actually *liked* being boys, if such a thing were possible.
2   It was generally agreed that, even if you didn't approve of violence, effeminate boys were "just asking to get beat up," and bore at least some of the responsibility when it happened.

I'd been fascinated by Antarctica since childhood. At first I just thought it was cool how weird and faraway and cold and different it was. I dreamed about exploring it myself, of how impressed everyone would be when I made it to the South Pole. But when puberty came along, my fascination with the polar regions shifted. The achievement of a difficult task lost its appeal, along with most of my other dreams, abandoned while my body transformed into a hideously ugly . . . *thing* I was apparently doomed to live inside of. But Antarctica still tempted me, this time with the idea of a vast, icy continent with nobody on it. A place where you could be, definitively, alone.

If I was in Antarctica, and walked out onto the ice, I could finally be sure that nobody was watching me. And, then, once I knew nobody could see me, I would . . . And my fantasy faded out from there. What I wanted was so dangerous that I couldn't risk even being seen thinking about it. I couldn't risk seeing myself think about it. But perhaps someday I would find a place where I could.

And so I found it incredibly significant when, while I was packing up for the first time I had ever moved entirely by myself, from my own place to my own place, NPR played a story about Antarctica. Apparently, someone or other had made some recordings of the wind blowing outside one of the polar stations. I don't know why those recordings were made or why, having been made, NPR decided to broadcast them to the world, but in that moment, it felt like it had been done just for me. I was going, I believed, to my Antarctica, the place where I could be alone. And there was another reason that moment felt so significant, which is the date that it happened. It was December 31, 1999, the end and beginning of everything.

The omen I saw in that NPR broadcast proved correct. The new apartment was everything I'd expected it to be. I found levels of solitude there that I've never achieved before or since, a near-perfect aloneness. But that apartment was like Antarctica in another way, too: it damn near killed me. That I survived is due to many factors, but the one that I always think of first is this: it was in that apartment that I first discovered the MTV show *Daria*, the greatest animated television show of all time.

Sick Sad World

To be clear, when I first started watching *Daria*, I didn't think it was the greatest. I didn't even think it was that good, necessarily. The characters seem[3] to be really broad, one-note, predictable. The writing is good, but not always as good as it thinks it is, and there are a handful of episodes that just don't work.

What drew me in at first was the character of Daria herself, with whom I identified to an almost painful degree. Smart, bookish, sarcastic, pessimistic. Embarrassed by her family, unpopular at school.[4] Hiding her extreme sensitivity behind a mask of sarcasm and condescension. She was the girl version of me,[5] and I couldn't get enough of her.

In particular, we were both always hiding. Daria seemed to share with me the same fear of visibility, of being exposed. Daria fears being caught showing emotion, because deep down she believes that she isn't valued in any of her relationships, so if she "forces" her emotions onto anyone, they'll turn away in disgust. When stressed, she retreats to her room, and it's worth noting that we rarely see her hanging out there with anyone else, in contrast to Jane's room, where we often see her with Daria, or a boyfriend.

---

3 Emphasis on "seem."
4 Though not as unpopular as she thinks she is.
5 So, me. As it turned out.

And this isolation doesn't seem like an accident; the decor in Daria's room is intentionally alienating,[6] helping ensure that she can be truly alone there, giving her a sanctuary where she can let down her guard, secure in the knowledge that any stray signs of emotion or vulnerability won't be used against her. Which I could understand because I'd already implemented the same strategy for myself: if you want to be alone, live somewhere unpleasant.

The apartment I was living in, and for which I'd agreed to pay $225 a month to the guy who worked at the used record store across the street, was just *fascinatingly* grim. The walls were cinder blocks painted yellowish brown, the color of a chain-smoker's teeth, a color that made everything in the apartment look vaguely ill.

It had one small window, facing north over the parking lot, which I looked out of only when I was expecting a pizza to be delivered, as the building had no outside doorbell. The shower was awkwardly crammed into a corner of the tiny bathroom, with a gap at the bottom from which mice would venture. It was my own version of Daria's padded walls and skeleton poster, and when I saw her bedroom, I knew I'd found a kindred spirit.

Student Life at the Dawn of the Millennium

Daria is a member of my generation, and a weird generation it is. I was born in 1979, which I think usually ends up as Gen X, but I've also seen categorized as millennial. It is neither; what it *is* is hard to pin down, but if you want to see it for yourself, watch *Daria*.

Take the season one episode "Cafe Disaffecto." When a local

---

6   It had padded walls because the previous residents had confined an insane person there. And the padding hadn't been removed before the house was sold. Look, just go with it.

cybercafe[7] is vandalized, Mr. O'Neill, the English teacher, is upset. Daria,[8] however, says that cybercafes are bad anyway because they don't offer real human connection. Which is of course ironic coming from Daria. She fears real human connection more than anything, and this comment is intended to shield her from it via ironic detachment.

But Mr. O'Neill is a huge sucker, and, taking Daria at her word, decides to open up an old-school coffeehouse, at which Daria winds up forced to read her work at an open mic night. Naturally, what she reads is not some personal expression of emotion, but rather a hard-boiled Tom Clancy–style techno-thriller, including either twenty or twenty-one dead communists.[9] It's a pitch-perfect character moment, the kind the show does so well. And the cybercafe/Cold War topics don't just pinpoint a moment in time; they show how that specific moment has shaped Daria, and me.

I think of my generation[10] as being distinguished by two factors: like millennials, we came to the internet at a young enough age for it to feel "natural" to us in a way that things we learn in adulthood never can. But unlike millennials, we can remember the Cold War.

Daria and I both spent our childhood aware that the world was locked in a struggle between the forces of Good (us) and Evil (the Soviets, or anyone else we didn't like). Then, just as we reached adolescence, that was suddenly no longer true. We were told that

---

7   alt.lawndale.com

8   Who shares with me the distinction of having been a classic honors student shit-stirrer.

9   "Twelve dead Russians, five dead Chinese, three or four dead Cubans."

10   Some people call it elder millennials—or geriatric millennials, a term that I find more upsetting than I'd like to admit.

we'd reached the End of History. The heroes had defeated the vil-
lains, the Big Boss at the end of the game. The story had reached
its ending.

Which was great news, we supposed, but it's a weird feeling to
be told that the story has ended, right at the age when you realize
that your own story is just beginning. The play ended just as we
were about to walk onstage. We were disoriented; the Cold War
had given us our elder peers' cynicism, and then its end had made
that cynicism pointless. *Daria*'s opening credits consist of Daria
refusing, Bartleby-like, to join in any group activity. That's what
first drew me to her: I, too, was sick of participating in the world,
and in this new apartment, I intended to stop.

You're Standing on My Neck

Lord knows, the world hadn't given me much incentive *not* to
withdraw. The last people I'd lived with had been during my
sophomore year, when I'd shared an apartment with three men,
all of whom were close friends with each other, and strangers to
me. This arrangement had come about because I didn't have any
friends at college who would room with me, so I got randomly
assigned to a group that needed a fourth person. They called me
"Mr. X," because they'd known nothing about me until we all
moved in, and they never really seemed interested in knowing
any more about me afterward.

Living with them had been awful, and I couldn't imagine anyone with whom it would be much better. In my three years at
college I had yet to make a single friend. My romantic life was
nonexistent: at the age of twenty-one I had never kissed anyone,
and the two or three dates I'd been on had ranged from "forgetta-
ble" to "disastrous."

I found it difficult to talk to my mother, and in general refused either to answer or return her phone calls.[11] My only pre-college friend still living nearby, and who currently filled the role of Amy's Official Crush, was out of town for the summer.

And then, a few months after I moved into the apartment of despair, my co-op job ended, leaving a full month before my fall classes would begin. So it seemed clear that it was time to finally put my plan into action, to finally live out my dream. It was time to go full recluse. I became largely nocturnal. My diet consisted of Kraft Mac & Cheese and Spicy Nacho Doritos, both of which I would buy in furtive trips to the gas station down the block at three in the morning, as well as the occasional pizza. Apart from those necessary excursions, I did . . . nothing much. Watch TV, eat junk food, jerk off, repeat.

I also took up smoking, because while suicide held no appeal for me, staying alive didn't seem that tempting, either. Complete solitude didn't feel as good as I'd imagined it would, and I didn't really have any other ideas. But it was during that isolation that I found *Daria*, much like Daria found Jane.

I Have Low Esteem for Everyone Else

The friendship between Daria and Jane is the central relationship in the show, and it starts in the first episode, on Daria's first day at Lawndale. She's been forced to take a "self-esteem" class, and there meets Jane, a cool, artsy student who's taken the class six times already. Both of them feel the same urge not to participate in such a transparently useless class (taught by the transparently useless Mr. O'Neill), but soon Daria decides to pull her good-student rank[12] to get them out of it.

---

11   My dad may have called me of his own volition at some point in my life, but offhand I can't remember it ever happening.

12   Jane only qualifies for that status in art class.

It's a deft setup that shows what their relationship is from the beginning. They're drawn to each other because they can see that neither will threaten the other one's self-protective detachment from the world. They both speak the same language, a classic adolescent pose of irony and condescension. But while that is the only language Daria can speak, Jane is a bit more socially adept. Daria, meanwhile, provides Jane an intellectual basis for her instinctive defense mechanisms, and the reassurance that she can't be dumb, as her teachers say, not if a smart person like Daria likes her.

Yet that shared defensiveness is a problem for their relationship, and it remains so throughout the series. Both of them resent each other a little bit, for compromising each other's independence. Jane is mean to Daria; not often, but regularly. And for much of the series, Daria doesn't do anything about it, because she believes Jane is the only friend she'll ever have. That codependency resonated with me; in retrospect, the main reason I didn't feel I had any friends at this time was because I refused to believe that I *could* make friends, and indeed that I already had done so. As had Daria. Watching the show, I saw my own relationship patterns in her. I could see how Jodie likes Daria, and considers them friends, while Daria likes (and admires, and envies) Jodie, but assumes that Jodie doesn't like her, couldn't possibly like her. Again, it's nothing obvious—subtle details in animation and line readings—which made it all the more effective at slipping into my psyche.

Your Shallowness Is So Thorough, It's Almost Like Depth

Of course, the show also shows just *why* Daria doesn't trust those around her. Every character in the show is a caricature of themselves. Quinn is shallow, Helen's a workaholic, Jake is hapless.

Mr. DeMartino and Ms. Barch are angry, Ms. Li is power-mad, Kevin and Brittany are dumb. I could never have connected with the show if it hadn't shown sympathy for why Daria is the way she is, the way I was. Like me, she was surrounded by people who were inexplicable, who it was impossible to imagine respecting. It took me a while to realize the way in which the show doesn't quite share Daria's opinion of them, doesn't actually agree that the characters are as ridiculous as Daria perceives them to be, and indeed as the show itself seems to portray them. But slowly I saw that, however ridiculous the characters might act, they all had their own motivations, their own reasons for what they did. Each of them are living out their own complex story.

The best example of this is the character of Brittany, a vapid blond cheerleader who talks in a high squeaky voice, and is . . . well, I wouldn't call her an intellectual. I found her almost repellent when I first started watching the show, but she grew to become one of my favorite characters, while still retaining every single quality that had annoyed me initially. We learn bits and pieces about her as the show goes along: her family is extremely wealthy, even by the high standards of Lawndale.[13] All the families in *Daria*, despite all being well-off, are delineated very precisely by economic status. Brittany's anxieties[14] are those of her status, and no other. She's not *that* smart, but she's smarter than she thinks she is. Certainly she's smarter than her boyfriend, and she knows that, more or less. She knows that she's too good for him, at least subconsciously. But he's the varsity quarterback, and the idea that she wouldn't want to date the QB just would never cross her mind. And, like so many people, she can't find the courage to risk being alone.

---

13 Her family lives in a gated community called Crewe Neck, whose motto is "Private and Proud." Not subtle, that motto.

14 I.e., whether she'll be judged for not having a Jacuzzi.

Brittany's guileless nature means she'll talk to anyone, so we get to see her in a wide variety of groups and situations. While she never stops being a caricature of a dumb blonde, you eventually realize that you've stopped caring. After all, most of us are pretty much caricatures, viewed in a certain light. But we are still people, still unique individuals with pasts and hopes and fears. What I learned from Brittany is that if I assess someone as ridiculous, as stupid, as one-dimensional and uninteresting, then I'd be making a mistake to write them off, even if my assessment was perfectly correct. Everyone, without exception, has hidden depths. And it's not just Brittany. I could tell more or less the same story about at least a dozen other characters on *Daria*, and that, ultimately, is why I believe that *Daria* helped save my life.

One effect of trauma is you learn to form quick first impressions as a matter of survival. And you get quite good at it! But what I didn't realize was that, when you act on those first impressions, and shut people out at the first sign of doubt, then you never get to see beyond them. *Daria* taught me that people I didn't understand nonetheless *understood themselves*, and that they had a whole story going on that might, if I learned it, allow me to know how to trust them, too.

That summer, I was in as dark a place as I'd ever been. The only strategy I'd come up with for dealing with the intolerable pain and fear I'd been feeling for years was to hide, to avoid all the people who I held responsible for causing that pain in the first place. And when that strategy failed, and I found myself in even more pain than I'd been in before, I had no backup plan. I simply could not see any way out of the pit I had dug for myself, until I came across Daria, sitting right there in the pit with me, caught in the same seemingly impossible dilemma.

Together, slowly, we were able to learn the vital truth we'd been missing. We learned that, even if we're right, even if the people around us are thoughtless, belligerent morons, always judging and attacking us for reasons that they themselves cannot comprehend, incapable of seeing the pain they cause us—even then, they are still just . . . people. Just like us. Even when they hurt us, the answer is not to shut them all out, because they are hurting, too, and the only way to deal with our pain is to help them deal with theirs. Trying to go it alone brought me to a dark place, and on my own, I couldn't find any way out. But, after that summer, I slowly began making my way back into the light, and I did it in part by following the boot prints left by a smart, sarcastic, sad cartoon teenager. If Daria could survive this Sick Sad World, then so could I.

CHAPTER TWELVE

Why Is There So Much Drama Around Bathrooms?

In 2016, Procter & Gamble released an advertisement for their brand of deodorant marketed for women, Secret. The plot of this thirty-second advertisement is this: three women[1] enter a public bathroom at what seems to be some kind of nightclub or performance venue. We then cut to a trans woman standing in the stall of that same bathroom. She's pacing back and forth, afraid to exit the stall.[2] Finally,[3] she takes a deep breath and goes out. As it cuts away to the Secret logo, you hear one of the cis girls casually saying, "Great dress," and another say, "It's really cute!" Tagline: "There's no wrong way to be a woman."

I can recite all this from memory because, in the months after it was released, I watched that commercial on YouTube hundreds

---

1 All of them are cis women, which of course you were already assuming when I said "women."

2 It's a big stall. All the characters, trans or cis, are coded as young, rich, beautiful, light-skinned.

3 Thanks to the confidence her deodorant has given her, we're presumably meant to think.

of times,[4] hoping that I, too, would someday get the courage[5] to walk into a women's bathroom. I could see the irony of getting comfort from a multinational conglomerate's[6] attempt to sell me deodorant. But there weren't many places I could go, many people out there who would tell me that I'd get there someday, that maybe, somehow, it was all going to be okay. This deodorant commercial would just have to do.

Eventually, on days when I was feeling convincingly feminine and I really had to pee, I started darting into women's bathrooms, at least in places that felt sufficiently anonymous.[7] But it was always terrifying. Every time, I expected to have my gender challenged by one of my restroom mates, after which a howling mob would materialize and march me to the town square while the villagers hurled rotten vegetables at me. Something like that. I would avoid all eye contact,[8] slip into my stall, and use it for its intended purpose.[9] Occasionally I would even linger in the stall in the hopes that other women would leave, just like in the first few seconds of that deodorant ad. But the howling mob never materialized.

One day I was at my favorite bar[10] with Jessie. I'd never been in the women's room there. It was small, and usually crowded, and since I'd been going to that bar for years, many of the regulars had known me pre-transition, at least enough to recognize me. I'd never sensed any discomfort from them, but I'd seen so

---

4   Literally, I'm not kidding, two hundred times at *least*.
5   Possibly from my deodorant.
6   Based in Cincinnati, no less!
7   Target, for example.
8   Though that's also standard protocol for men's rooms, so it came naturally.
9   Peeing. I didn't *poop* in public bathrooms, I'm not some kind of animal.
10 The Heart and Dagger, but don't show up there and ruin it for me.

much hysteria online about trans people in bathrooms, I still half-expected whenever I entered one that I would trigger some kind of emergency Klaxon, or someone would point at me and shout, "*J'accuse!*" So when I had to pee, I told Jessie that I'd need to go home to do so. Without hesitation, she told me that was silly, took my arm, and marched me over to the women's room. I peed without incident,[11] and felt a huge wave of relief. That bar had long been my second home, and it now felt like it still could be, even as a woman.

And before too long, I didn't just lose my fear of women's rooms. I fell in love with them. Especially women's rooms in bars. That deodorant ad was right about this much: women compliment each other's clothes in bathrooms. Women's rooms in bars might be cramped, dingy, with not enough stalls to meet the needs of a bunch of drunk women who need to urinate, hurl, do drugs, or finger each other.[12] But they are almost always still filled with good vibes,[13] camaraderie, mutual support. The graffiti in women's bathrooms is dramatically different than in men's, filled with messages of empowerment. "Don't let him tell you you're not worth it." "You deserve better." "Pussy is power." "<Man's Name> is a rapist, stay away from him." Any woman in a bathroom who needs something: toilet paper, a tampon, advice, compliments; whatever it is will be provided, by any woman who's able. They're safe hideaways within the dangerous jungle of mixed-gender nightlife.

Even as I got comfortable with restrooms, there was still one place that frightened me: the gym locker room. After all, in a

---

11   I don't remember if I did a key bump while I was in there, but if not it would have been one of the rare times I used that bathroom with Jessie and didn't do any blow.

12   If they've been doing MDMA.

13   And, admittedly, bad smells.

women's room, you still have privacy in your stalls.[14] But in the locker room, people get naked. And I felt a lingering cloud of guilt over the fact that, yes, I am attracted to women. I enjoy seeing female bodies, and attractive ones are disproportionately represented in locker rooms. I couldn't deny that the prospect of seeing some nude athletic female bodies appealed to me, and I didn't want to be a perv. But I also wanted to get some exercise back into my routine, and one day a coworker convinced me to join her fancy gym downtown. "Sure it's expensive, but it's super fancy and luxurious, so you'll be able to convince yourself that it's a treat, and actually go there. Better to have an expensive gym membership you use than a cheap one you don't." And since I'd be going there with her, and thus she could help extricate me, should the denizens of the locker room rise as one, emitting howls of rage and disgust—and since the bathroom forays seemed to have worked out fine—I agreed.

I never got fully naked where anyone could see me. Or even slightly naked. The showers had little changing rooms attached, with a bench and hooks and so on, so I would get fully dressed in there after my shower, even though I wasn't fully dry, and all my clothes would cling weirdly to me in the steam. But I didn't want anyone to see my penis, because that seemed to me something somebody could legitimately be upset by, without necessarily being transphobic.[15] So I got in and out of there as fast as I could. Yes, there were naked women in there, and yes, they came into my field of vision sometimes, and yes, I often found them quite attractive. How could I not? But I stayed nonchalant and kept

---

14   Unlike with men's rooms and their urinals, or even more horrifyingly, their troughs. There's a reason there's less camaraderie in men's rooms, it's because we all have to get our dicks out in front of each other. Whose idea was this?

15   Also I just didn't want anyone to see my penis. I'm not a fan of it!

tight control of my gaze at all times. I'm confident that I never made anybody feel uncomfortable. I would have spent even less time in that locker room, but I still felt it was impossible for me to appear in public without a full face of makeup, so I had to stay long enough to apply it, dutifully keeping my eyes fixed on my reflection in the mirror.

Then one day after our workout, a friend[16] was ready to leave, and I told her I couldn't, I still had to put my makeup on.

"You know, Amy, you don't really need it, you look pretty without it."

I looked in the mirror, and I was shocked to realize that she was right. I didn't need the makeup anymore! That was a perfectly good-looking woman looking back at me from the mirror, without any artificial enhancements.[17] And so while I never got as comfortable in locker rooms as I am in bathrooms,[18] I still remember that as a place where I got one of the many great gifts I've been given by other women in those single-gender refuges over the years.

Not long ago,[19] something happened that I'd long been expecting, and dreading: TERFs found my Twitter account. I'd tried

---

16   When I'd convinced her to join the gym with me, I didn't realize how in shape she was. Then she asked me to join a spin class with her. I went in the class excited at sharing a possible bonding experience with a friend who was starting to become a close friend. I left the class feeling like I was dying, and in fact might already be dead. We still went to the gym together, but from then on we worked out separately, in ways fitted to our respective abilities.

17   I mean, that's not really true. I was taking hormones every day, as I will for the rest of my life, and I'd had to get hair transplants to alleviate the receding hairline I'd started to develop before figuring out my gender. But without makeup, anyway.

18   Partly of course because I haven't spent as much time in them. I haven't been to the gym since COVID. I want to get back in the exercise habit, but also, I don't *really* want to. Exercise is hard!

19   Meaning in 2023. I'm doing a time jump here.

to avoid anything too "political," anything that might cause some notable transphobe out there to unleash their herd of flying monkeys on me. But I had found myself unable to stay silent on the issue of a certain billionaire YA author and her attacks on people like me. And when you say anything about the TERFs' queen, they will all come crawling out of the woodwork to defend her honor.

They weren't saying anything I haven't heard a million times, of course, nothing I hadn't said to myself when I was struggling with my identity. And certainly they weren't the first people to tweet mean things at me. But it's one thing to have evangelical Christians calling me evil, or well-intentioned parents quoting misleading *New York Times* articles, or just straight-up assholes calling me a freak and mocking my appearance. It hurts a lot more when I am accused of hating women, of undermining and endangering them, of not caring about keeping them safe. I would never endanger women! I love women! And the whole point I'm trying to make is, I *am* a woman! If I'm putting women in danger, then I'm putting myself in danger. Why would I want to do that?

But, of course, they claim that I'm not a woman.[20] The phrase "man in a dress" gets thrown around a lot.[21] That's the same image I was given of trans people my whole life, the same epithet I hurled at myself over and over again in the years before and after my transition, one that I still find entering my mind in times of doubt. I've worked so hard to overcome that self-hatred, and now I have

---

20   Although, still, they can't deny that I think, or at least claim, that I'm a woman. That's the whole thing they're upset about. But that still means that it would be nonsensical for me to endanger women!

21   Actually, what I saw more was "bloke in a dress," since in my experience TERFs are disproportionately British.

people echoing it back at me, using the same phrases, the same lines of attack, that cost me decades of closeted agony. It just sucks.

And the whole trans issue is just not that hard! All you have to do is believe me, take me at my word when I tell you that I'm a woman. After that, the solution to all these "controversies" out there will seem trivial. Should I be allowed in a women's bathroom? Well, duh. I'm a woman. And it's not hard to believe me, either. Did you see me on TV? What did you see? If you saw anything other than "a woman," then I have to tell you, you're in a small minority. There are so many grandpas from Middle America that, without even trying, accepted me as a woman, and gendered me correctly. It's only difficult if you decide to be difficult about it.

It's been years since I started "infiltrating" single-gender spaces, and nothing bad has ever happened. I love those spaces as much as any cis woman, and possibly more so, having spent my life without them. And I guess there are two questions I have for the people that want to keep me out.

First: If men want to dress up like women and sneak into women's bathrooms . . . can't they do that anyway? Couldn't they always do that? What do trans people have to do with this? Why bring us into it?

And second: You know this dystopia you describe? Where "men in dresses" are free to "invade" the sanctity of women's restrooms? That's been the reality in the Bay Area for over a decade. If it's so dangerous to women, why has nothing happened yet? Why aren't the cis women living in this "nightmare" you imagine clamoring to change it? They've learned the simple yet revelatory truth. Trans people are just people. They need to pee sometimes. Let them.

CHAPTER THIRTEEN

Why Do You Do Tarot?

I consider myself an atheist, although I tend to avoid saying so. And when I do, I always want to immediately follow up by saying that I'm not what you might call an "evangelical atheist." I hope I'm not smug about my atheism, and I certainly don't believe that people with religious faith are less intelligent than I am. But I cannot bring myself to believe in the basic concept of a god, by which I mean a being that has an identity, that has free will and independent opinions, and that is more or less responsible for everything that goes on in the world.

I know that for many religious people, the belief that there are one or more gods, and that they are watching over us, and have a plan for our lives and the world as a whole, is vital. And they find it hard to imagine getting through their day-to-day lives without the comfort of their religious faith. But to me, the idea that there is someone responsible for everything seems unbearable. I can deal with all the bullshit of the world, the evil and pain and injustice, and accept the bad that is inextricably mixed with all the joys of life. But to think that all of it is *intentional*, that it was somebody's *idea*, I find harder to take.

I know the exact moment my journey to atheism began: my First Communion. I was raised in an entirely Catholic environment, and believed in its doctrines unquestioningly, the same way I believed in everything adults told me. Then, in second grade, as we prepared to take our First Communion, I learned about the Catholic dogma of transubstantiation,[1] which meant that, during the Liturgy of the Eucharist, the bread and wine on the altar actually, literally, got transformed into Jesus's body and blood. So, when we had our test run the day before our actual First Communion, we were just receiving bread and wine, even if a priest was handing it to us, but the next day we'd be getting the body and blood of Jesus himself. I was fascinated to experience this transformation myself, but when the moment came . . . it was just bread and wine, same as the day before. I'd had nosebleeds, I knew what blood tasted like, that metallic tang. This was just wine, cloying and sweet. So, I'd learned something new: sometimes church stuff wasn't real.

I'd learned already that sometimes adults had rules for an actual reason, and sometimes they had rules just because they were rules.[2] I'd assumed that church rules were in that first category, because God had imposed them, and he probably knew what he was doing. But once I'd found out that some church stuff was just made up by grown-ups, I started asking myself about the rest of them.

It was still a long journey to atheism. I tried my best to find satisfying solutions to the problems I encountered. I found heaven and hell difficult to get my head around.[3] Heaven seemed boring.

---

1 What a thrillingly long word!

2 For example, rules about what clothes you could wear, or which parts of your body you were allowed to ask questions about.

3 When I was nine or ten years old, my mom and I had an epic argument over a homework assignment I had for religion class, in which I was supposed to de-

And hell was supposed to be eternal, but like, doesn't God win at the end? And when he does, is he just going to keep hell running? And if not, didn't that mean we'd all get to heaven eventually, no matter how much we sinned?[4]

In middle school, when I went to that Seventh-day Adventist summer camp, I met people who believed in more or less the same God as me, but had a much different idea about it. The man who led our Bible study group once explained that anyone who didn't believe in Jesus would go to hell, even someone like Gandhi. That seemed horribly cruel to me, but at the same time, I found that viewpoint to make more sense, in a way. Why did we Catholics have to go to mass every Sunday if non-Catholics got to get to heaven anyway just by being good? But really, I just couldn't make the afterlife make sense. The idea that God would condemn people, people who he had created, who lived in a world he had created, to an eternity of suffering and torment, just because they hadn't followed a particular set of rules that, to be frank, he had done a pretty poor job of communicating? That idea was intolerable to me.

So, by the time I got my driver's license, I stopped going to church. Like most teenagers, I'd started sleeping in a lot more, so it wasn't implausible to tell my parents that I was going to the 6:00 p.m. mass at St. Joe's downtown, rather than the 9:30 a.m. one at our local parish. Then at 5:45 or so, I would get in my car, drive randomly around Dayton for an hour, and head back home.[5] As time went on, my lies about the subject became more and more transparent. I claimed to attend mass at the university

---

scribe what heaven was like. Which, I argued, was unknowable! Nobody who'd been there had reported back!

4   Like, even bad sins, like saying bad words, or wanting to know how sex worked.

5   Catholic mass is quite a reliable length, fortunately.

chapel, but when my mom would ask me which priest had offici-
ated I would say, "Oh, I don't really know which one is which."
After I graduated, and had been living in my own apartment for a
year, my mother asked which parish I attended and I just said that
I didn't know what its name was. What was she going to do about
it? Did she *want* to know that I wasn't Catholic anymore? And that
I was unlikely ever to return?

I can also date the moment my journey to atheism was completed: Election Day, 2000. That night, for the last time in my
life, I prayed, and really made a sincere attempt to believe that my
prayer was being heard. But whether my prayer was heard or not,
it didn't change the fact that Theresa LePore had fucked up the
ballot design in Palm Beach County, and so this evil idiot failson
was going to become our president. And that was pretty much it
for me, as far as believing in God goes.[6] I now considered myself an atheist, no longer believing that there was any will or consciousness in the universe except for our own.[7]

And yet, I do tarot, and at times[8] I base decisions on the result
of tarot readings, the random selection of cards from a deck, even
though I genuinely think it's random. And not just tarot; I've dab-
bled in astrology, and it's a similar vibe. I don't believe in astrology,
and yet at the same time, I would be hesitant to date a Pisces.[9]

This hobby dates back to before my apostate days; I got in
trouble for fortune-telling once as a fairly young child. I was play-

---

6   I would rather believe that there was nobody in charge than that there was
somebody in charge and they were fine with the idea of President George W.
Bush.

7   Assuming that we have a will and consciousness ourselves, which at times I
doubted.

8   While writing this essay, for example.

9   Sorry, Pisces! I'm willing to be proved wrong.

ing a game with my brother where I grabbed some random Scrabble tiles and threw them on the ground, then explained to him what they meant. I'd say something like "Well, you see the *P* lying on top of that *G*, that means that your pride is getting in the way of your . . . um . . . good nature." Or whatever, it was all improvised, but we were having fun with it. But when my mom saw us doing it, she made me stop, in the half-ashamed way she did on those occasions when she was enforcing a rule she didn't 100 percent believe in. Our parish was sensitive to the occult; many of the families regarded Halloween as a satanic ritual, and on October 31 would throw a "King's Kids" party as an alternative. All the kids would still get to dress in costumes, and at some point in the proceedings they would go around to all the adults and get candy. But, like, in a Christian way, I guess?

My mom had no patience for that,[10] nor for the extreme censorship practiced by some of my relatives, who considered *Star Wars* immoral because "the Force" was essentially demonic. But she still felt like she couldn't let me go around interpreting messages from the Scrabble gods.

Of course, I didn't believe in the Scrabble gods then, and I don't believe in any gods now. So how can I justify doing tarot? If I can't believe in the transubstantiation, how can I justify seeking wisdom from a deck of cards made up by some Victorian nutjobs?[11] Well, for one thing: tarot works. If I'm going to be a cold, practical materialist, then I have to accept the evidence of my own lived experience. And over and over, I have seen tarot readings help people. I have been given tarot readings that have absolutely

---

10   We attended the King's Kids party once or twice, since all the people we knew were there, but my mom made it clear that Halloween was just fine as far as she was concerned.

11   Or Edwardian, maybe, they were right on the cusp.

helped me, have been positive influences in my life. And when I've given tarot readings, even to complete strangers, the *vast* majority of the time, the person receiving the reading has found it extremely helpful.[12] Until I experienced it for myself, I always vaguely assumed that tarot was for suckers, in much the way I still believe that psychics are.[13] That anybody who got something out of a tarot reading was just the victim of a scammer with an understanding of how to manipulate people.

But I'm not a scammer, and I don't think I have any special skill at manipulating people.[14] So, when I've given a tarot reading, and the person I gave it to found it successful, there must be something else going on. But that "something else" doesn't have to be magic. Let's assume that this deck of cards has no supernatural qualities, that it is just a stack of seventy-eight pieces of paper. Or not paper exactly? I don't know what you call the material a playing card is made from. Anyway, just a stack of cards made in some factory, with variously colored inks applied in different places. How, then, can they provide such value?

The answer, it seems to me, is that in the same way that a wall is made up of bricks, or a body is made up of cells, or a book is made up of words, our mind is made up of stories. Story is the fundamental unit of our mental existence. If you want to get in your mind and mess around with it, then you need a tool that's shaped like it, a tool that's shaped like a story. And a tarot deck, it turns out, is quite good at churning out different shapes of story.[15]

---

12   Or at least they've said as much, but I think I have a decent sense of when people are sincere, and when they're just saying something to be polite.

13   Which is hypocritical of me, but I'm fine with that.

14   I'm sure I'm as manipulative as anyone else, but I don't *practice* it.

15   Though I don't want to overstate the importance of story; it's not ideal for, say, deciding who should sit on the Iron Throne and rule the Seven Kingdoms.

There are two types of cards in the tarot deck, the Major and Minor Arcana, and they correspond to the two main types of stories in the world. There are stories as in "I read a story in the newspaper," and stories as in *The Odyssey*.

The Minor Arcana tell the stories that fill our day-to-day lives. They tell stories of having a drink with a loved one, feeling sad in the rain, working in your garden, heading out on a trip, playing a game with your friends. They repeat themselves, in the same way the headlines do. The four suits deal with our bodies, our spirits, our emotions, and our deeds.

The Major Arcana tell one big story, the Hero's Journey type of epic, the sorts of stories we find inspiration in, that feel larger than ourselves, larger than our times. They exist in the same heightened reality as *Romeo and Juliet*, and they speak to those parts of ourselves that want to be part of the epic poem of human existence. We don't have easy access to our minds. We don't often realize that; intuitively, surely if something exists in our minds, we could access it whenever we wanted? But of course that's not the case. If it were, people would use Google way less. Think of how you use Google yourself: sometimes, yes, it's to find information unknown to you. But most of the time, aren't you using it to look up something that you "do" know, but can't bring to your conscious mind? Something that you will recognize as already having been present in your brain; indeed, if for whatever reason your search terms threw up several different possible answers to your question, you'll immediately know which one is the one you sought. There are parts of our brain we're aware of, but have trouble finding. There are no direct flights from our consciousness

---

15, *cont'd*   (I'm told not everyone will recognize this as a *Game of Thrones* reference, but I can't write a whole book without complaining about the ending of *Game of Thrones* somewhere, I'm only human.)

to that part of our brain, so they need to be routed through our physical perception.

We might need to draw out a diagram, or write out the steps of our long division. Maybe we're trying to remember the name of a song, and we need to sing it out loud in the hopes that we will eventually come to the lyric that contains the title. And maybe, if we filter our thoughts through the images and symbols devised by some eccentric English people a century ago, we'll be able to reach new parts of ourselves. The tarot cards do not create those parts of ourselves. How could they? But what they can do is point the way, guide our consciousness to unknown destinations. After all, none of us is all that special. Our minds all have a similar terrain. And so the signs that have guided others along their journey can probably guide us as well.

But I should come clean: I *really* got into tarot to meet women.[16] I was in a weird place in 2017, simultaneously the best and worst time of my life. On the plus side, I knew what gender I was, and being a woman in the world was an exhilarating, joyous experience. I had spent my entire life avoiding conversation with anyone I wasn't extremely intimate with; now I *loved* talking to strangers. I loved trading outfit compliments in a women's room. I loved chatting with respectfully flirtatious Uber drivers. I loved striking up a conversation with a random Canadian woman in the Minneapolis airport while we both waited out a layover at adjoining gates.

But I was single, and had just divorced the only person I had ever had a real romantic relationship with, and who I had also delegated my entire social life to for the previous decade. So I had few friends, and that lifetime of avoiding conversation had made chatting with strangers difficult, if enjoyably so.

---

16   Well, people in general, but mainly women.

The thing is, I don't have much patience for small talk. I'm happy to do the "So, what do you do?" "Did you grow up in the Bay Area?" "What brings you here?" thing for a minute or two, but if I had my druthers, I'd then go straight to "What are your dreams?" "How's your relationship with your mother?" "What traumas are you still healing from?" Other people can find this alarming, so I usually manage to restrain myself, but at the cost of feeling a bit lost at times for what to talk about.

So for someone like me, tarot was a perfect icebreaker. I developed a habit of going to my favorite bar most evenings, sitting by myself on the back patio, and setting a tarot deck out in front of me while I nursed a sauvignon blanc. And fairly often, before too long, somebody would come up. "Is that a tarot deck?" "Yup." "So, do you read tarot?" "I do! Would you like a reading?" In my experience, nobody ever says no to that question. Who doesn't want to learn more about themselves, about their future? Bullshit or not, it's appealing.

I was able on a regular basis to have conversations that were intense in a way I found interesting, while only lasting five or ten minutes, leaving both sides without any obligations when they ended. I would tell them they had recently completed a spiritual or creative journey, and they would tell me what they had been passionate about of late. Or I would tell them that they were filled with emotions, but weren't sure which ones to act on, and they would tell me about their hopes and fears. And so on, and so on, and what continually amazed me is that, almost always, they would say something like "Oh, wow, that's definitely true." "I really feel that." "That was super helpful." "I really needed to hear that."

People would learn something from my readings, would come away changed. It was just a deck of cards, which had been shuffled, selected, and interpreted by a random drunk bitch (me). And yet, something profound would seem to take place. Usually, I wouldn't even know what it was; the other person would say that it had been meaningful, but not explain why, and it wasn't really my business. It was between them and themselves. I had just been a bridge between them, an extension of the tarot deck itself.

To be part of a profound experience, one that you help create but don't fully understand, and to do so over and over again, has an impact. And to have that happen while transitioning, seeing the world transform from a prison to a paradise, from a world of suffering and lies to one of love and joy and truth, made the bitter, cramped version of atheism I'd been moving toward seem untenable. I had to face the truth: things were happening that I couldn't explain. I made peace with the mystery. After all, physics itself proves that physics can't explain everything.

If you're interested, I encourage you to give it a try. To do a reading, first you need to pick out a spread, an arrangement of cards, each of which will have a specific meaning. There are a million of them out there, but my tendency is to just do three cards, and apply whatever dialectic makes sense in the moment. When preparing this chapter, I chose "Intention, Expectation, Reality," meaning I decided that (moving from left to right) the first card would represent what I was consciously, intentionally attempting to achieve with my writing, the second card would represent what I was subconsciously expecting or hoping to achieve with it, and the third card would represent what the writing that resulted from the tension of those first two cards would embody. But the basic format can be applied to whatever situation you find yourself in.

For example:

"L: What I would gain from accepting this job offer. M: What I would gain from rejecting this job offer. R: The goal I have in making my decision."

"L: Why do I want to do x? M: Why do I fear doing x? R: Why am I considering x in the first place?"

"L: My mother. M: My father. R: Myself."

"L: Past. M: Future. R: Present."

Then you need to know what the cards themselves mean. Or at least, I find it helpful. As I've said, there are two types of cards, Major and Minor. You may find one or the other more difficult; the Major because each card is in a way independent, and must be understood in its own way; and the Minor because there are more of them, and the discrete meaning of each must be pulled forth from the din of its surrounding context. Though the Major Arcana have a context of their own, which is crucial, otherwise you wouldn't be able to make too much sense of them, they would be dangling antecedents. It would take a whole book[17] to explain the Major Arcana, but here's my attempt to sum up each card in 240 characters or less:

0. The Fool

The Fool is numbered *0*, but really it has no number, it is the bodyless, intent-less, outside protagonist of the story. It is the person asking the question, but it is also all of us, blithely strolling through a universe we don't remotely understand.

1. The Magician

Okay, we're trying to figure out the universe. We know some of it is unknowable, up there, that god/spirit/fantasy/virtue/emotion stuff, and we know that some of it is all too knowable, down there in the ground/body/hunger/sensation/physics stuff, and we want to grasp both sides.

---

17 Specifically, *Seventy-Eight Degrees of Wisdom*, which is where I learned tarot.

2. The High Priestess

We've realized that we need access to a higher plane than we've been granted. We tried to grasp both sides, and we failed, and we need someone who has already grasped both sides to show us the mysteries they have learned, and tell us we are worthy of doing the same.

3. The Empress

Someone who has achieved what we want. She understands everything, and she rests in the garden of herself, which lives its own life according to her will, because her will is the nature of all things. To know all is to rule all.

4. The Emperor

Someone who has the power to achieve even without understanding. Whatever the emperor wishes will be reality. He will only fail to achieve what he doesn't choose to pursue.

5. The Hierophant

Now we grow beyond ourselves. The hierophant, the high priest, has done this, *this very tarot*, before you, and knows it better than you ever will. The hierophant is the pope, who rules above kings, or says that he does, in any case.

6. The Lovers

We are not alone in the world. We do not even exist without others. We cannot be ourselves except with another. Love is vital because without it we cannot know ourselves. Our thoughts, our minds, our souls, do not exist until they have been shared, been joined in communion.

7. The Chariot

Success. We have done it. Now what? We show ourselves to others, we show the thing we have done, we parade it before the world, and they are happy, and we are happy. And then?

8. Strength

So what then is true strength? It is not the strength to overpower, it is the strength to persuade, the strength that allows a woman to live at peace with a lion. The strength to live in accordance with the universe, not the strength that attempts to shape it to our will.

9. The Hermit

We need more wisdom than we started with. The hermit is the one who has the wisdom that we seek, but he is also ourselves, seeking that wisdom in solitude, withdrawing from the world's illusions to find the truth within ourselves.

10. The Wheel of Fortune

Give up the illusion of control. The universe has its own plans for us, and will reward and punish us in its own way, in its own time. We must learn how to exist whether the times are good or bad, because both will come.

11. Justice

And yet within the whims of fate, right and wrong still exist. We must apply our own justice, and we must be prepared for others to do the same to us. True justice can never be achieved, but injustice must be fought, nonetheless.

12. The Hanged Man

Suspended by one foot, the hanged man is powerless, upside down, yet serene. He teaches us that in the spins of the Wheel, in the face of injustice, our own happiness is always and only within our own power.

13. Death

All things end. Death is not just about our own death, or the deaths of any person. It is about the finiteness of everything, and that everything we have we will someday lose, and yet must continue.

14. Temperance

Everything exists in balance, in tension. We will never find a single answer, but must always find an equilibrium, a balance, the blend of earth and water, body and mind, joy and suffering, light and dark. Be ready to flow.

15. The Devil

Wisdom and insight are all well and good, but we live in bodies. The figures in the Devil echo those of the lovers, chained to the prosaic world of hunger, sleep, sex, and pain. But are they unhappy?

16. The Tower

Death taught us that everything ends. The Tower teaches us that no matter how wise we are, sometimes catastrophe will strike. Violent disruption, war, disaster—sometimes no amount of inner peace can protect us.

17. The Star

Yet now we reach the true destination of our quest. Passing through the catastrophe we emerge into a strange world, and find the true insight, the truths that transcend words, that can only be learned, not taught.

18. The Moon

The Star is the wonder at the wisdom we have found, the Moon is the strangeness of it. In the light of the moon, things appear in strange forms. What's that lobster doing? You must find your own answer.

19. The Sun

And when we have these moments, where we have passed through the veil, and learned the deep truths, we will be filled with light, with joy, the radiance of the sun. It's time to shine that light back in the world.

20. Judgment

What now? What do we make of the world now that we know its truths? This is the moment when we must use the wisdom we learned, and judge the world, and ourselves. It is exhilarating, terrifying, and inescapable.

21. The World

And so we go on, out into the world we started in and never left. The Sun, the Moon, the Star, these are places we can visit, but we are ever and always called back to the place we started, and the people we share it with.

That was exhausting! And let's be honest, most of the time, we're not trying to pierce the veil, see the light behind the moon, perceive the mysteries of the universe. Most days aren't High Holy Days, most desires aren't spiritual, most questions aren't eternal. In the Major Arcana, we slip the surly bonds of earth, and touch the face of God. But in the Minor Arcana, each card represents the patterns and randomness of life as we actually live it. But they are given a loose order by their suit and number. The suits break up our experience into the four elements of the ancient Greeks.

Cups are water, emotion, feeling. Happy, sad, angry, jealous, joyful, giggly, worried, excited, focused, dazed, hurt, sympathetic, wistful. They are the tides that wash through us, the currents that carry our boat where they will.

Pentacles are earth, physicality, body, money, hunger, shelter, sex, comfort, pain, lust, sleep, illness, experience. They are the real, physical world we live in; they are the animal bodies we inhabit, our monkey brains that want to live and play and interact in the world, the joys and hopes and fears that require no faith or wisdom or insight. The pain of getting smacked in the face, the pleasure of pulling on a cozy sweater, the taste of a cool glass of water on a hot day.

Wands are fire, spirit, creativity, intuition, faith, philosophy, charisma. They are the parts of us that leap up out of the physical world, unbounded by physics, by reason. They connect us to what is beyond us, what exists above and below, before and after, the truths that exist outside of our bodies and emotions.

Swords are air, action, impact, movement. They are our ability to change the world—to decide, act, move, have an effect. They are weapons, tools, toys, they create change.

Each ace has the same design, a hand reaching down from the sky, representing that each of the suits are a gift to us from the universe, a birthright, something we don't have to earn.

The pages represent our first attempts to wield the suit, the knights represent our mature use of it, the queens represent full understanding of the suit, and the kings represent the full control of it. What the rest of the cards mean is largely up to you. For that matter, what every card means is up to you. They're just cards!

By the way, when I pulled those cards I mentioned earlier, here's what I saw:

The Three of Pentacles showed up in the first spot, the one representing what I was attempting to achieve in this piece. In it we see a craftsman carving three pentacles into the ornamental crown of a pillar in a church. To me, that represents an attempt to blend the physical aspect of pentacles with a higher purpose. In this case, it means that I'm trying to write something that will sell books, while also offering some enlightenment to the reader.

The Lovers showed up in the second spot, the one representing what I was hoping to achieve. Being Major, it represents something deeper, more integral to who I am. In this context, I'm somebody that believes in the power of human connection, and what I truly hope to accomplish is to help people connect with each other, to be able to share love with more people than you could before, through what I'm writing here about tarot.

The High Priestess showed up in the third spot, representing what I am capable of achieving here. As so often is the case, the card representing reality is something of a letdown. The High Priestess decides who is eligible to continue their journey to wisdom. I fear that I'm not worthy, that there are deeper meanings than I understand. I know that there are truths that I understand but have failed to communicate in this piece.

And that may be the case! But nonetheless, that reading gave me what I needed. It brought some hopes and fears to the surface, where I could see them, accept them, and move on. It wasn't anything I already knew, but nonetheless, before I looked at those three random cards, I felt blocked, and afterward, the words started to flow.

Plus, if you look at this piece, you can see that I more or less shaped it around those three cards: starting with the earthen mundanity of my atheism, and the tarot deck itself, then moving to the human connection that tarot has brought me, and then discussing the magic behind it all, and laying down my instructions for performing the ritual. Was it magic? I don't know, but it allowed me to finish this piece and send it off to my editor, and as far as he's concerned, that will be miraculous enough.

CHAPTER FOURTEEN

Why in God's Name Did They Make
Cartoon All-Stars to the Rescue?

I was born in 1979. My wife was born in 1996. One thing that I've learned from this is that you simply cannot prepare someone who was born in 1996 for the experience of watching the 1990 animated special *Cartoon All-Stars to the Rescue (CASttR)*. No matter what you tell them, they will not believe it could be as bonkers as you say it is. They have to see it for themselves to understand, to the extent that it is understandable at all.

For those of you who weren't born yet, I should explain that, on April 21, 1990, every major television network in the United States dropped their usual programming for thirty minutes, and instead they all simultaneously broadcast an animated special, funded by McDonald's, about the danger of doing drugs. And I can't speak for everyone else, but I can confirm that there was at least one ten-year-old in Dayton, Ohio, who was *stoked* for it.

I wasn't stoked for it because I was excited for an anti-drug message. At ten years old, I'd already had plenty of anti-drug messages, and the pace would not slow down in the years to come. I suppose they had worked, in the sense that if someone had offered me, a ten-year-old child, drugs, I would have said no.[1] Nonetheless, I had already worn out my appetite for that rhetoric. Once they had stated that drugs were bad, there wasn't anywhere else for them to go. But apparently McDonald's had other ideas.

I wasn't even stoked at the ostensible selling point of the special, the chance to see "all my favorite cartoon characters together," not really. Even ten-year-olds are sophisticated-enough media consumers to understand that a cameo by your favorite character in an anti-drug special was unlikely to give you what you were looking for out of that character. I was looking forward to the special simply because it was a special.

As a child I was obsessed with "significant" moments; just a few months earlier I had been almost unbearably excited for New Year's. It was going to be 1990! A whole new decade! With an extra *9* in it! I was allowed to stay up until midnight, and spent the time between midnight and sleep noting to myself all my "firsts" of the decade.[2] So I was excited for this special simply because it was a "first" of its own; I'd read in the papers that never before had every major network agreed to air the same show at the same time. Which I believed because I read the TV listings cover to cover every week, and I'd certainly never seen anything like it.

Upon reflection, it seems kind of strange that I read the TV listings so diligently, given that we didn't have cable, and even in terms of network shows, there were very few I was allowed to watch. But I get why I did it. One of my favorite movies around then was *Short Circuit*, the Steve Guttenberg–Ally Sheedy comedy

---

1   Spoiler: this abstinence would not last forever.

2   The only one I still recall: my first food of the 1990s was a Dorito. An actual, brand-name Dorito, not the ones that we got at the Metro Market that came in a plain white bag labeled "Nacho Cheese Tortilla Chips," with no other branding.

about a military robot that develops sentience.[3] There's a scene shortly after the robot comes to life where he shows up at Ally Sheedy's house and demands, "Input!" She gives him a dictionary or whatever, he flips through reading the entire thing in a few seconds, throws it aside, and then again demands, "Input!"

That spoke to me because that was the same way I felt about the world. I wanted more information, and I didn't really much care what information it was. At the library, I'd read all the books titled things like *500 Craziest Sports Facts!*, which, while lacking in plot, character development, or the remotest relevance to my life, fulfilled my main criteria of just having a lot of *stuff* to learn. Years later, playing Trivial Pursuit, I baffled my friends with my ability to come up with the answer Paavo Nurmi.[4] But I could, because I'd come across it during this period of indiscriminate "Input!"

So the TV listings[5] were catnip for me, just pure unadulterated input. And that Sunday, as soon as the paper came, I eagerly flipped forward to Saturday morning, and there it was:

2 (NBC): *Cartoon All-Stars to the Rescue*
7 (CBS): *Cartoon All-Stars to the Rescue*
16 (PBS): <some bullshit>
22 (ABC): *Cartoon All-Stars to the Rescue*
45 (FOX): *Cartoon All-Stars to the Rescue*

---

3   I had no idea that Fisher Stevens was in brownface, not that that would have seemed remarkable or problematic to me at the time.
4   A Finnish distance runner from the 1920s, not that there's any remotely conceivable reason you should know that.
5   No, I do *not* mean *TV Guide*; why would we pay for a *TV Guide* when the listings were included for free in the Sunday *Dayton Daily News*? Having a *TV Guide* was a sign of middle-class opulence; they were only found in the sort of houses that got mad at you for wearing shoes on the couch.

And so what I was looking forward to that Saturday, as much as anything else, was the opportunity to flip between all four stations and see the same show playing. Nobody had ever been able to do that before! And that was really all I cared about. Sure, I probably told my parents something along the lines of "I think it's really great that it will have this anti-drug message," because grown-ups had made it *extremely* clear that they wanted me to think drugs were bad, and pleasing grown-ups was one of my main strengths.[6] But I don't think I had any particular thoughts of my own on drugs; having solved the riddle of what grown-ups wanted me to say about them,[7] I didn't really feel like there was anything more to interest me on the subject. And yet it was around this age that I had my first drug experience.

This isn't a shocking story, along the lines of the youthful drug experiences of *Go Ask Alice*.[8] It was just an accident. All I remember is that I was sick, and somehow or other I accidentally took a double dose of Triaminic, and soon thereafter I felt *weird*, but in a good way. At one point, I was lying on the carpet looking at our living room's popcorn ceiling, fascinated by the patterns I was seeing, and I thought to myself, *I am* really *enjoying looking at this ceiling!* and being fascinated by the fact that I could get such enjoyment out of something so mundane. Just by taking some extra medicine!

I didn't think that I was having a drug experience at the time. I knew Triaminic couldn't be a drug because drugs were called

---

6   I'm still pretty good at it.

7   Namely: "Scary! Bad! No!"

8   If you're not familiar, *Go Ask Alice* is a fake "diary" of a nonexistent teenage drug addict, fabricated by a Mormon youth counselor in 1971, which became wildly popular for its "authentic" look at drug use in adolescence. I once had a small role in a stage adaptation.

"drugs," and you could only get them from evil strangers in alleys. But in retrospect, it was the beginning of my interest in experimenting with chemical alterations, an interest I've had ever since. And even the best efforts of McDonald's, George H. W. and Barbara Bush, the Senate Judiciary Committee, Oscar-winning actor George C. Scott, and "All your favorite cartoon characters!" never really had a chance to dissuade me from it. Why?

Well, the first-order answer to that question is: because their TV special fucking sucked. It was doomed from the start, even apart from the anti-drug messaging. The problem with bringing "all your favorite cartoon characters" together is that cartoon characters don't all live in the same universe, or have the same sensibilities. Even ten-year-old Amy could see that Winnie the Pooh and Slimer shouldn't be in the same cartoon. Yet in *CASttR*, not only do they appear together in the very first scene, but Alvin and the Chipmunks join them. *And* Garfield. *And* Baby Kermit![9] And Alf! Who, and I cannot stress this enough, *was not a cartoon*!

As an aside, I had no issue at the time with the inclusion of Alf per se. I was a big fan. One of my early memories is of a Monday night when Ronald Reagan made a national broadcast (given how old I think I was, it was probably either about Iran-Contra or the bombing of Libya), thus preempting *Alf*. I was *furious*. Was this the start of my disillusionment with the Republican Party? Possibly! Though there were other factors.

One of those other factors, getting back to the point, was the War on Drugs, whose surreal futility and cluelessness were so clearly on display in *CASttR*. And yet this cartoon was seriously intended, by serious people, to actually change the way that drugs were used in America. The Senate Judiciary Committee received

---

9   *Baby* Kermit!

a prerelease screening of this show.[10] Again, this show's plot sum-
mary on Wikipedia contains the following phrases:

> *"Smoke," an anthropomorphic cloud of smoke*
> *He is hesitant until one of them steals his wallet*
> *A police officer, who is revealed to be Bugs Bunny*
> *He and Smoke chase after her, until they fall down a manhole and*
> *   meet up with Michelangelo*
> *Baby Kermit, Baby Miss Piggy, and Baby Gonzo take him on a tour*
> *   of the human brain*
> *He looks at himself in a small mirror and is shocked to see Alf*
> *They go talk to their parents about his drug addiction*

And then, after seeing all of that play out on-screen, the chair-
man of the Senate Judiciary Committee[11] said, "The most powerful weapon that we know in politics is the cartoon and we hope
that the cartoon will be the most powerful tool to educate our
children." And that he wanted to watch it again when it aired!

Sure, he may have been lying,[12] but even if you set that aside,
every major television network agreed to sacrifice a half hour
of advertising revenue, and countless apparently well-meaning
schools and nonprofit organizations helped promote it to as many
people as possible. And while it's hard to believe, the only plausible explanation I can come up with is that all these people *actually
thought this special would help solve a real problem*. Which is baffling!

When I encounter a conundrum like this, when I realize that
a large group of reasonable people believe something that I find
self-evidently nonsensical, I know that there's a lesson for me

---

10   Makes you want to go into politics, doesn't it?

11   One Joseph R. Biden Jr.

12   God, I hope so.

there, an opportunity to expand my mind and learn something new. These people must have had their reasons for believing in *CASttR*, and the broader War on Drugs. And if I write them off as "crazy" for that belief, I'm missing the chance to learn a new perspective on the world.

I don't think I'm blowing anyone's mind when I say that the War on Drugs was misguided. But it's still the case that, even as marijuana has transformed from "horrifying gateway drug, will almost certainly kill you" to "Our app makes cannabis delivery easy!," the mentality that created the War on Drugs is still present in our society. Many people who welcome marijuana decriminalization nonetheless feel, perhaps subconsciously, that there is still some category of "bad drugs," and that fighting those "bad drugs" is still the government's moral obligation.

I disagree. I think the entire concept of a War on Drugs was a mistake from day one. I have used, and continue to use, a wide variety of drugs, and while they've certainly had their downsides, I don't regret them. It hasn't kept me from being successful. I believe that drugs have been good for me. Without them I might not be here today. Maybe that's not true, the '80s kid in me is still kind of always expecting it all to catch up to me, and I'm not ruling out the possibility that it will someday. But the lived experience of actual drug users doesn't get a lot of exposure in our society, so I feel a responsibility to share mine, unfinished as my story may be. If nothing else, I just want to do my part to make sure that Papa Smurf never again shares a scene with Baby Gonzo.

CHAPTER FIFTEEN

Okay Then, So What Have Your Experiences with Drugs Been Like?

As Isaac Newton was developing the science of optics, one of the experiments he performed was to jam a blunt needle into his eye socket, use it to distort the shape of his eyeball in various ways, and record the results. He wanted to understand the universe, and he realized that to do so, he first had to understand himself, his own perceptions. He wanted to understand how his eyes processed visual information, so that he could "reverse engineer" it, and have a fuller understanding of reality outside of himself, outside of the filtered version of reality that he could experience directly.

Drugs, to me, are taking that experiment to the next level. How is all the input that is being received by *me*, by the "I" that is doing the thinking,[1] how is that input being altered by my brain, before I ever receive it? And what can I learn from those alterations? I always want to try on new perspectives, if only to better understand my own. And getting high, more or less by definition,

---

1 The *sum* that *cogito*, to butcher Descartes's famous phrase.

gives you a new perspective to experiment with. But it took me a while to make that discovery.[2]

There were some people who drank at my high school, but not many, and as far as I was aware, none of my classmates smoked weed. It later proved that this wasn't quite as true as I thought. Some people in my circle smoked weed occasionally, they just never looped me in on that fact, presumably because they thought I'd be uninterested, or even frightened. I hope it wasn't because they thought I'd narc on them, but in any case, they clearly misunderstood me.

Yes, I tended to follow the rules. But I only did so because, well, they were the rules and following them seemed to please various authority figures. I didn't think the rules had much value in themselves, I just knew that following them would make adults well-disposed toward me, which could come in handy in all sorts of ways. Breaking a rule never bothered me particularly, as long as I'd considered the consequences of doing so.

So, on Graduation Weekend my senior year, when I was invited to a party at Rosie's drinking-age sister's house, at which it had been made very clear that alcoholic beverages would be widely available, I didn't hesitate to accept. Our social circle was made up of honors students,[3] and I sensed we were all thinking more or less the same thing, which was that we had followed the rules all through high school, and had achieved the goal they'd laid out for us of graduating high school and being accepted to a good college, and since the point of the "no drinking" rule was that drinking might prevent us from achieving that goal, there was

---

2   And yes, I know that there are downsides, I'm not pretending there's no danger in them, or that addiction hasn't ruined countless lives. Again, my analogy is to Newton sticking a *needle in his fucking eyeball.*

3   Albeit the honors students from unstable households.

no longer any real need to take it seriously. We played along with your game; we did what you wanted. Now we needed a drink.

In yet another amusing-in-retrospect moment of gender expression, my first alcoholic beverage was a strawberry wine cooler. It was fine. I was happy to be trying this whole "alcohol" thing out, but I wasn't trying to make a big thing out of it like some of us were. Two of my friends[4] had been telling everyone how they were buying a case of wine, and they were going to drink all of it and get *so wasted*. But when the time came, they only managed to get partway through the second bottle before becoming drunk to the point of incoherence, writhing around on the floor, in a way that might have been erotic, as people do when they're raised in a sex-negative environment and need to lose control of their bodies before they can experience them.

While I was happy to be breaking the rules, getting drunk was slightly frightening. I only knew what it looked like from the outside, that it caused people to do what they normally wouldn't in a given context. Alcohol seemed to hinder people's ability to read the room, and reading the room had always felt crucial to me. I didn't see any reason behind most social conventions and expectations, so I couldn't predict what unwritten rules might suddenly crop up. The only way to avoid slipups with potentially disastrous results was constant vigilance, constant observation. I always felt in danger of being called out and humiliated for breaking a rule that everyone else understood implicitly, but which I'd never been taught.

So actually experiencing drunkenness for myself had to wait until a few months later when I went to college. The University of

---

4 Who were also my partners in a developing love triangle that would span decades.

Dayton, whatever its other flaws, was absolutely first-rate when it came to exposing its students to alcohol, at least when I was there. I'd joined the marching band without much thinking about it, band having provided much of my social network in high school. I would come to regret it,[5] but the few weeks I lasted before I quit provided a couple advantages.

One advantage was that I got to move in a week early, as band practice started before the semester did. The other was that all the freshmen in the band were assigned a "big brother/sister." One of the responsibilities that role entailed was escorting their assigned freshmen to their first off-campus rager, which, while not quite officially sanctioned, took place on a night that was prominently left blank on the first week's schedule, and everyone clearly knew why.

Different regions have different names for the combination of fruit punch and Everclear that fuels many of the most serious college parties. At Dayton we called it "hairy buff." That night I learned the reason for its popularity: it gets you drunk even if you weren't intending to. I didn't even realize I was drunk until I'd walked back to my dorm, laid down in my bunk, and tried to read a book. The words were swimming in front of my eyes, and I realized: *This is it! I'm actually drunk! This is what it feels like!*

I found this just as fascinating as my Triaminic trip all those years ago. My reservations about losing control remained, but now I saw the upside, and it wasn't the anxiety suppression, or the loosening of social inhibitions. It was simply the experience of having a different brain for a while.[6]

---

5   Like so many decisions I was making around then.
6   Since my Triaminic experience, it has always seemed obvious to me that there must be some middle ground between addiction and abstinence, even if Barbara Bush, McGruff the Crime Dog, and Baby Miss Piggy didn't see it.

And just like we can enhance the weather inside our homes to suit us better than "natural" weather, we can also enhance the chemistry inside our brains to suit us better than the "natural" chemistry. That's not particularly controversial. Certainly many of the people working so hard on behalf of the War on Drugs had no problem with caffeine, psychiatric medications, and various other chemical alterations that somehow didn't count as "drugs." One example that's particularly relevant to me: I will be on the mind-altering drug estrogen for the rest of my life.[7]

And so the problem with the propaganda produced by the War on Drugs was that it refused to acknowledge this broadly understood fact. The thing I'd always wondered was this: If drugs were as bad as all that, why did anyone do them in the first place? Peer pressure didn't seem all that convincing, when weighed against the supposedly dire consequences of drug use. Surely there must be another reason? And now, as I lay in my dorm giggling, unable to see the words on a page in front of me, I had my answer.

I still didn't drink much during college.[8] In part that was due to a fairly bleak social life, in a place that I fit into about as badly as I'd ever fit in anywhere. But partly (and relatedly) it was because however much I enjoyed playing with my brain this way, it didn't negate my fear of losing control of myself in a social setting, and making the fatal error that would cause my few friends to turn away in revulsion and leave me utterly alone. And the behavior of my fellow students was not much of an advertisement for the benefits of alcohol. I once woke up at 10:00 a.m. on St. Patrick's Day[9] to find someone passed out on the grass outside my window. Still,

---

7 I certainly hope so, anyway.
8 At least, not by prevailing local norms.
9 A *huge* party day in Dayton, as well as much of the Eastern and Central time zones, as my California-born wife was shocked to discover.

I certainly became familiar with the contours of drunkenness, and was glad to add that to my store of knowledge.

But that was just alcohol. My first actual experience with what *CASttR* meant by "drugs" didn't take place until I was twenty-five. I'd been in a community theater production of *The Taming of the Shrew*, which was easily the worst play I was ever part of.[10] But I'd become friends with Kelly, who played the titular shrew and shared my love of Radiohead.

Like a few others in the cast, she was in a college theater program, and only in this low-rent community theater production because they were dissatisfied with the parts they played at school. As we left the theater one night, Kelly and Eli, one of her classmates, called me over to Eli's car. As I walked over, Kelly asked me, "Do you smoke?" In my innocence, I thought she meant cigarettes, so I replied, "Sometimes?"

She pulled me into the back seat, and Eli started packing a small glass pipe with what I realized was marijuana. I would have died rather than admit my misunderstanding, or that this was my first time smoking "pot," as the kids call it. I nonetheless fumbled with the pipe, but Kelly bailed me out by asking, "Oh, have you never used a one-hitter before?" I said I hadn't,[11] and she showed me how to hold a thumb over the carb before taking a drag. How did it feel? Not really like anything, in retrospect. I think I was still using it wrong and didn't inhale much smoke. What it felt like mainly was relief: I hadn't been exposed. Kelly and Eli still thought I was cool.

The next night they proved as much by inviting me to a party. A college party among the theater students at a middling Midwestern university is an interesting experience, everyone there

---

10   If any of my castmates read this: Sorry! I like you as people!
11   Not technically a lie.

being conscious of themselves as both part of an elite (artists in a world of accountants and car salesmen) and an underclass (artists in general, and in particular artists who didn't have the talent, connections, or money to get into a top-tier program, and who therefore quite rightly suspected they had vanishingly small chances at ever successfully making a career out of their art). That mix creates people with a great deal of interest in mind-altering chemicals, whether to quiet the fear, expand their creative horizons, or demonstrate to the dentists and HR managers of the world that they didn't believe in their rules and parameters. At this first party, numbness appeared to be the priority, which gave me the opportunity to try my second illegal drug. This time it was Vicodin,[12] crushed into a powder to be snorted. I felt comfortable admitting that I'd never done this before, because I'd never even *heard* of anyone doing it. And I can't say I recommend it; it doesn't do much beyond potentially damaging your liver. But these were the drug experts, as far as I was concerned, so if they said it would be fun I wasn't about to argue.

A few days later, Kelly and I started dating, and would eventually get married. We were the same in so many ways, among them that we'd both been subjected to the War on Drugs in a conservative Catholic context. We were still prone to believe the horror stories, and to project them onto other people's experiences. So when Kelly declared that "we" didn't do cocaine, because she'd seen a friend have some real bad comedowns from it, and that "we" didn't do psychedelics because she'd once babysat a group of friends on a shrooms trip that had seemed unpleasant to her, I went along with it.

---

12 So not technically illegal, I guess, though I have to assume that using cash to buy prescription drugs from a guy named "Bones" might not have been a *completely* legitimate transaction.

I still found security in following clearly defined rules, and if those were the rules Kelly wanted, I was happy to go along. But after we moved to the Bay Area, and started hanging out with the stand-up and theater crowds out here, we started being friends with people who did do cocaine and psychedelics. And as far as I could see, they seemed to enjoy it in about the same way people enjoyed drinking. Like, some of them were taking it too far and kind of knew it and regretted it, and while many saw it as a guilty pleasure, an indulgence, the majority enjoyed it, and were glad it was part of their lives.

I felt a vague regret that I'd agreed not to experiment, but I didn't really think about it much. The whole reason I'd outsourced most of my decision-making to Kelly was so that I didn't have to think about it. But then Kelly left, I realized I was trans, and everything I believed about myself had to be reexamined. And when I did, I realized that I'd been living my whole life blind to what my mind was capable of. And it seemed to me that it was time to get back to experimenting. Time to try some new shapes for my brain, and see which ones were better.

Two of the first post-divorce friends I made were Mara and Jessie. Both were at least tangentially related to the local stand-up scene, which means it's not shocking that each of them were also drug users.[13] The one that I was most interested in, because it seemed to be the most common and most social, was cocaine. Jessie was quite happy to give me that opportunity.

I don't remember the first time I did coke, although I'm sure it was with her. I do know that I didn't make any pretense of having done it before. I'd learned that people who wouldn't accept me weren't people I should have in my life. And at this time in par-

---

13   One occasional, one more dedicated.

ticular, I'd become intoxicated by being truthful about myself, by telling people about how I felt, and had always felt, about gender, about Kelly, about my own value, about everything.

That first time with cocaine wasn't that memorable. Cocaine changes your perspective less than even alcohol, let alone marijuana. It's less helpful for providing new insights. For me, it's always been like a more powerful and reliable caffeine. My first experiences were much like that first college party, a sudden realization that I was excited about what I was doing, eager to share my thoughts, and also that my jaw hurt because I was grinding my teeth. The latter was a familiar situation, but had always been due to stress. It was strange to feel that sensation while also being happy.

Cocaine did help me expand my horizons, but in a noticeably different way. Alcohol and marijuana worked on a mental level; they made my brain respond differently and more intensely to whatever inputs they might be getting (conversation, television, my own introspection, internet porn), so that, even when I was sober, I was able to appreciate those experiences more directly. But that internal side wasn't where cocaine showed its value that much. It was still helpful, but largely as an add-on to those other chemicals. It made me more motivated to follow the new lines of thought that they provoked, and crucially, to write those thoughts down, and to make plans for what I would *do* about them.

Where it shone was not in giving me new thoughts, but new experiences. Alcohol lowered my inhibitions, but also sapped my energy. Cocaine fixed that problem and then some, and gave me options I'd never had before. After-after parties. Dancing in the kitchen with my coke dealer as his girlfriend eyed me suspiciously. Meeting people who had lived different lives than me, with nothing connecting us besides our shared appreciation for cocaine and adventure. Hooking up on vacation. Watching the sun rise from the downtown Marriott, accompanied by one friend and six strangers, all of us wearing hotel bathrobes. There's a fascinating nocturnal[14] world out there, and the only people who can access it are those who have done some blow.

Cocaine also served for a few years as the medical treatment I didn't know I needed for ADD. I spent the late 2010s rebuilding my entire life, and that wouldn't have been possible without *something* treating my ADD. Actual medication intended for ADD has been far more effective, unsurprisingly. But compared to having no treatment for it whatsoever, cocaine was a godsend.

I had some fun cocaine experiences with Mara as well, and she was somebody who proved the point that, as with alcohol, there was a whole class of people who used and enjoyed cocaine occasionally, as an entertainment option, without needing or even wanting to do it more often. It was also Mara who introduced me to psychedelics. She was a cautious person, which I found comforting, since, by all accounts, taking hallucinogens could be pretty intense, in a way that was apparently indescribable. Nearly everyone who tried seemed to end up admitting that their description was a pale imitation of the reality. I would just have to experience it for myself.

We took some acid in a diner at the top of the Castro, and by the time our check came, the lights were beginning to sparkle in a way I found fascinating. We went to a dance club and danced in the middle of a crowd of gay men. At one point[15] I realized that my life had changed in a way that would never be undone, that everything before now had been prologue, a decades-long hibernation, and that was over forever. Later we went up and sat on one

---

14   And very-early-diurnal.
15   It was while they were playing "Sweet Dreams (Are Made of This)."

of the benches at the top of Dolores Park and looked out over the San Francisco skyline, and spent an hour or so just laughing and laughing, unable to believe how wonderful everything was. And we were right. Everything is that wonderful. I can't always feel it the way I did then, but thanks to LSD, I always know that it's true.

A few months later, we did shrooms. We went to China Beach, on one of the only days I've ever experienced actual beach weather in San Francisco. Mara had to go help our friend to the bathroom right when my trip started, and I was frightened for a minute. But I lay down and looked up at the rocky cliff stretching into the sky above me, and I saw it breathing, and I breathed with it. I realized that what I'd come to learn was how to be comfortable being alone. And that being alone didn't mean being unsupported. I was connected to everything, and I didn't need other people mediating that connection for me. And I've been more comfortable being alone ever since.

After they came back, we hung out and marveled at the crystal of the sun shining off the waves, and the cliffs across the Bay, and the way birds had whole lives they lived apart from us, and again, we laughed and laughed. Mara and I are no longer in contact with each other, and our parting was not amicable. But when I think of her, I think of this: the two of us together, at the dawn of my true, real life, with everything possible, laughing and laughing at how full and deep and wonderful the universe was, and had always been.

CHAPTER SIXTEEN

So If You're Trans, Does That Mean You Like Guys?

The first time I realized something had changed was during Pride Weekend 2019. My two friends and I had just finished the Dyke March, and they'd gone in a bar together to use the bathroom. I was still pretty wired from the bumps we'd done before the march, and I'd been sexually harassed by some of the bar's founders years earlier,[1] so I stayed outside to smoke. Some guy bummed a cigarette off me, and when he started a conversation, it took some time for me to realize that he was flirting with me. And then I realized something shocking: I liked it! I started flirting back.[2] When we'd finished our cigarettes, he said to me, "You wanna go over there and make out?"

Now, I'd never before in my life been attracted to a man. And I had tried to be! I'd always known I was unmanly, and tried to conceal my femininity. But back then I never thought that I was a woman, because I didn't know that was something a person could

---

1 Pre-transition, if that matters.
2 Which is much easier to do as a girl; all you have to do is smile and look vaguely interested.

think. The only reason I knew of for my inability to feel like a man was to be gay. So I tried jerking off to *Romeo and Juliet*–era Leonardo DiCaprio.[3] But it didn't work. I simply didn't like boys.

And until this moment, I'd thought that was still true. But when he asked the question, I felt a little shiver. He was cute! In kind of a skeezy way, to be sure. A couple inches taller than me, reasonably fit, seemingly in his mid-twenties, stubble hovering on the border between intentional and lazy. And, almost involuntarily, I heard myself answering, "Yeah, sure." Smiling, looking away, performing indifference in that coy way that makes it clear that you're not indifferent at all.

We walked down the block to a doorway, which didn't seem any more desirable as a make-out spot than the corner we'd been on, but he seemed to be the expert here, so I deferred to his judgment. I wasn't sure what to do, but even these days, when I know how to make the first move, I never do. I have no interest in boys who don't know what they want, and how to get it.

He slipped an arm around my waist and pulled me against his body (it was so hard! So different, so . . . non-curvy, so lacking in grace and subtlety!) and bent down to my willingly upturned face, and began making out; deep, hard, so hard I knew I'd feel it in my lips for a little while. His stubble scratched against me, and to my amazement I wasn't disgusted by it, even though facial hair had always repelled me. His arm kept me tight against him, pressing up against my skimpy outfit with his T-shirt and jeans.

I could smell him, the sweat of a man out in the city summer sun, and I could feel him, feel his cock stiffening against my hip. And I liked it! I liked how it felt. I realized: I had opinions about cocks! I was more into some cocks than others, apparently, and his

---

3   Who, in terms of manliness, wasn't exactly a lumberjack.

felt like one I'd be into. And even as I thought it, he pulled away for a second, and said something I'd never imagined hearing: "I want to stick my cock in your mouth."

The feeling was so intense I could barely function. Even now, when I think about it, I get an echo of it, a feeling unlike any I'd ever had with a woman. It was an electric jolt, a thrill that ran down the front of my body, from my neck down to my crotch. Ten minutes earlier, I'd never imagined having sex with a man, but as that shock ran through me, I didn't want anything else. "I'd like that, too," I said. Not even shyly this time, but in a throaty voice that made my lust undeniable.

"Do you have a place around here?"

Hmm. A dilemma. I would later become much more comfortable with letting strange men into my home for sex purposes, but at that moment I was only ten minutes into my bisexuality, and that seemed like a bridge too far. Plus, I lived in Oakland, so it was also *literally* a bridge too far.

"Well, I don't know, what about you? You got anywhere we can go?"

I forget how he answered, but he deflected in a way that made it clear that his home was off the table. I couldn't be sure, but the impression I got was that (a) he had roommates, and (b) said roommates would make fun of him for hooking up with a tranny.[4]

---

4 Among the things I identify as is this: I am a tranny. Before we settled on *trans women*, people like me were given all kinds of different identifiers, and out of all of them, *tranny* is the most fun to say. *Tranny* also fits easily into conversational English, unlike most alternatives. *Trans* is an adjective, not a noun, and we can't make it a noun because it ends in 's' but isn't plural. *LGBT* is a speed bump in the middle of every sentence that uses it, and always seems to be either too specific, or not specific enough. (Which is why it keeps mutating into *LGBTQ, LGBTQIA, LGBTQ+, LGBTQQIP2SA,* etc.) *Queer* is solid, but too hazy and undefined for some usages. *Shemale* is just embarrassing for everyone involved. *Trans* originated

But the offer of a cock was one I badly wanted to accept. We resumed making out, for lack of a better idea, and Mara and Jessie reemerged. I was a bit embarrassed at the fact that they'd found me, their avowedly lesbian friend, immediately after completing Dyke March, making out with a semi-sketchy dude she'd just met.

I introduced them; I don't remember if I knew his name at that point, but if I ever learned it, I forgot it within the afternoon. So let's call him Jared, after early-2000s Jared Leto, whom he vaguely resembled, in terms of demeanor and overall kempt-ness. I explained that yes, I'd been making out with him and wanted to continue doing so, but we didn't have a place. So we decided to continue with our original plan, which was to go to El Rio for the huge dance party that always followed Dyke March.

Taking a guy to a lesbian dance party was an odd move, but Jared and I both believed that if we stuck together, an opportunity to have sex would materialize somehow. I was open to ditching

---

as a prefix, but all the words formed with that prefix, such as *transvestite, transgender, transexual,* even *trans*,* are all now considered derogatory to some degree. Of course, *tranny* is also seen as derogatory, but there's a difference. Words like *transvestite* are considered derogatory due to the nature of the words themselves. All of them are attempts to capture the experience of trans-ness etymologically, and all of them fail to do so. Whereas, if *tranny* had never existed as an epithet, and we were where we are now, terminology-wise (i.e., seemingly settled on adjectival, stand-alone *trans* as the standard neutral identifier) then nobody would have a problem with people making it into a noun as *tranny.* It's only considered insulting because it originated at a time where the identity *itself* was an insult. It was considered insulting to describe someone as believing themselves to be female despite having a male body. Extremely insulting, in fact! The sort of insult that people get violent over, that gets people killed. And yet, that "insult" describes me precisely. *Tranny,* when it was coined, meant precisely the same thing that *trans woman* means now. So, since I don't actually consider it an insult to be described in that way, I see no reason why I shouldn't call myself a tranny. It just means that I'm a woman in a male body, same as it's always meant. Being a tranny shouldn't have been shameful then, and I refuse to be shamed by it now.

my friends for this dude when that opportunity arose, but until then we'd all stick together.

Amazingly enough, as soon as we'd decided this, Mara saw someone getting out of a taxi—an actual, honest-to-god San Francisco taxi, which had been a rare enough sight even before Uber came along, and now felt miraculous. Jessie got in front and I took the middle seat in back, between Mara and Jared. The taxi driver was like a taxi driver in a movie, living up to the cliché of the garrulous, street-smart, experienced taxi driver who dispenses wisdom and points out the sites of obscure events in local history.

But it was wasted on me, because I had my hand in Jared's lap and was surreptitiously[5] rubbing his hard cock through his jeans. He was doing his best not to react, but he couldn't stop the occasional gasp or twitch, and he couldn't get away. I was discovering my favorite thing about boys, which is how easy they can be to control when they're horny.

Things kind of petered out from there; the line at El Rio was too long, so we went to the bar next door. Jared and I kept trying to talk each other into taking us back to their place, but it became clear that neither of us would budge, and at some point we wound up getting separated. I didn't make too much of an effort to find him again, and I doubt he did, either. I'd gotten all I was going to get out of him, and it had been amazing. I was in a euphoric daze, floating on air.[6] A part of me had woken up that I never knew existed, and I couldn't stop marveling over it, feeling the texture of it in my mind, imagining the possible experiences that now might be available to me.

I later learned that it is quite common for trans people, after

---

5 Although probably not *that* surreptitiously.
6 And also, by this point, fairly drunk.

transitioning, to experience changes in their sexual orientation, whether or not they go on hormones.[7] This seems a bit counter-intuitive, but, speaking from my own experience, I can see in hindsight why it happened. For one, my feelings toward men had never been "Eh, they don't really do anything for me," but rather an intense revulsion and disgust. I could see now that much, if not all, of that reaction was a side effect of the hatred I felt for my own body. The characteristics that most repelled me about men were the same ones that I felt the most dysphoria about, body and facial hair being chief among them (pre-transition I was quite hirsute, and hated it). Once that self-hatred ebbed and I could see men without seeing myself in them, my feelings turned out to be different.

Another thing, and the one that I was most conscious of in the moment, was the validation they could offer me as a woman. It's not as if transitioning had gotten rid of my dysphoria, my self-hatred. It had gotten a lot better in the preceding couple of years, but nonetheless, much of the time I still saw myself as awkward and ugly, somebody who other people would always see as a man trying to look like a woman. Who could never pass, never be thought of as a woman without the accompanying "trans."

As supportive as my female friends were,[8] my dysphoria still insisted that any compliments they were giving me were just because they liked me as a person. Which was nice and all, but didn't make me feel any better about myself as a body, as a woman. Jared, however, clearly didn't give a shit about me as a person. He was motivated purely by lust. I was just an object to him, and for the first time in my entire life, I knew what it was like to *feel* desired.

---

7   Just one of the many delightful surprises of transitioning!

8   And Jessie was supportive to the point of having sex with me occasionally.

I can see now that I'd *been* desired plenty of times. There was one woman who I had a decade-long sexual-tension-friendship with, but who I never really thought was interested in me. Then, quite recently, I was cleaning out my apartment and I found some letters that she had written to me in a box of memorabilia. And when I read them, it was *incredibly obvious* that she was into me. I just couldn't see it at the time. I liked her a lot, but I didn't pick up on the dynamic. And when I think back now on other women I've known, I can see that at least some of them were attracted to me as well. But I'd never *believed* that they were, due to the potent combination of gender dysphoria and religious shame that had defined my life thus far. And now it turned out that I'd apparently been waiting my whole life to (a) be a woman, and (b) have some loser dude try to stick his dick in me.

After a life fixated on girl drama, the nuances of who was mad at who, and why, and what position every other woman they had ever met had taken on the dispute, I'd found a ticket back into boy world—a world where relationships were transactional, where all you had to worry about was whether you got what you wanted, and it didn't even occur to you that anyone else might be affected. A world where I could have a relationship that was only based on one part of myself, where nobody asked or cared about my childhood traumas, my mental health history, or even my name in some cases.[9]

Where people could just make out without having to then negotiate how to explain it to anyone else, without caring how it affected their make-out partner's other relationships. Without

---

9   Well, certainly nobody cared. They ask occasionally, but it's clearly because they heard somewhere that chicks like that kind of thing? When you ask them questions? Sometimes if you ask a chick questions she'll suck your dick afterward?

caring how it affected their make-out partner, really. Where lust was an understandable motivation, not a mortal sin. I hadn't gotten to go all the way with Jared, but in that moment I knew that, before long, some lucky guy was going to get to stick his dick in my mouth, and that it would make me feel good in a way that nothing else ever had.

It was a great day, and the first of a whole category of memories I'll treasure: the nameless guy who did blow off my tits in a Toronto Airbnb. The guy who left his laptop playing that night's *Jimmy Kimmel Live!* the whole time we had sex. The guy who thought he was being smooth by drawing a picture of me, only to produce an image so hideous that I blocked his number. A threesome with two twenty-year-olds who kept nervously checking in with each other to make sure they weren't gay. That afternoon with Jared, I'd grown up in a way, entered a new phase of my life, and it's one that I've enjoyed. But I have to admit: I didn't exactly exemplify the spirit of the Dyke March.

CHAPTER SEVENTEEN

How Did You and Genevieve Meet?

Coming out as trans means being faced with a whole new set of semantic questions, not all of which can be answered. If I tell a story that is about myself when I was in the Boy Scouts, what pronoun do I use to describe the lead character? If I say "he," that implies that I was a boy back then, whereas I believe myself to have been female from birth. But how can I claim that I was a girl back then, when whoever I was back then would have sincerely and emphatically denied such a claim?

Cis people sometimes have trouble understanding the double-exposure way that trans people see the past, the way I can look back at a child who would say, "I am definitely a boy," and believe that child to be both wrong and right. So, if I say "she," then it'll take them out of the story, make them think it's some performative trans nonsense, instead of just a story from my childhood. I guess I'll say "he." But in the spring of 2021, I had a different semantic question to struggle with: Am I in my first lesbian relationship?

Genevieve and I met because Ben, her brother, was dating my friend Jessie. Genevieve had just moved home and Ben was annoying her, so she asked Jessie to come over and distract him.

Jessie was over at my place and didn't have a car, so Genevieve had to pick her up, and she came up to chill in my apartment for a bit. I'd never believed in love at first sight, but I was smitten as soon as she walked in. Her dark eyes with their thick lashes, her smile, her hoodie, her comfort in an unfamiliar situation, the way she was so unmistakably an Oakland girl. And we connected in conversation. I gave her a tarot reading, my preferred way to form a connection with a cute stranger. After a while she and Jessie left for her mom's house, but maybe an hour later she texted me and asked if she could come back and spend the night. Of course I agreed immediately; my only concern was not to seem *too* enthusiastic. She was just asking to crash on my couch, but nonetheless it was already clear that we had a real connection. There was only one problem: she was straight.

There's a joke in *The Hitchhiker's Guide to the Galaxy* about the many verb tenses needed to describe events in a universe with time travel, and I often think about that when I struggle with these semantic problems. After all, transitioning is a kind of time travel, or maybe a trip between dimensions. When I became a woman, I acquired a different history, one where I was in Girl Scouts and learned about makeup and so on, and it doesn't feel fabricated. It feels like that whole life was going on in parallel with mine, and transitioning is what merged them together. So when I married Kelly, I was actually a woman myself. And surely a marriage between two women consitutes a lesbian relationship? Except, it was also 2006 in Ohio, so it was legally impossible for two women to be married—it was in the state constitution and everything. Although, if you think about it . . . Well, you can think about it for a long time, and never reach the end.

Early on in our friendship, the way I understood it was this: Genevieve had learned she was straight when she was in college. She went to Wells, a tiny school that had been a women's college until 2004, and was thus still a pretty gay environment. She started sleeping with one of her friends, but it ended badly. She didn't go into too much detail, but the (now ex-) friend told her, basically, that she was clearly straight because she'd failed at being a lesbian. She lost that friend, along with some others in the group, and decided it was time to go back to dating men. Surely, then, that was her first lesbian relationship? But she doesn't think so, and I think I agree. If she thought of herself as straight the whole time, can it have been a lesbian relationship any more than my marriage was? Or any less?

And so it turns out, it's not just trans people (or time travelers, for that matter) who run into these semantic walls. If something is assumed to be unchangeable, and then changes, the language can break down. Whether it's the forward arrow of time, a person's gender, their sexuality, or anything else, changing it can shake loose the precise definition of words (*she, he, before, after, husband, wife*) that are so fundamental to the structure of language that changing their meanings requires whole new sentence structures that don't yet exist. Eventually they will. I would imagine that a serf who, through some bizarre circumstance, attained a title of nobility would have difficulty talking about his past; not out of embarrassment, but because a noble saying, "When I was a serf" would be describing a situation that was socially impossible, but had nonetheless happened.

Semantic problems are not "merely" semantic. All the knowledge we share with each other, we share through language. And thus, the precise meanings of words within that language, and who gets to define them, underpins the whole fabric of everything we know. If a physicist makes a particular measurement, they may learn something true, entirely independent of any semantic meaning. But if the physicist is not able to communicate what they've learned, the measurement benefits nobody. Dismissing a debate by saying "you're just arguing over semantics" is missing the point, because *every* debate is arguing over semantics. Semantics are all we have.

The first time Genevieve and I kissed was while I was helping her break up with her boyfriend. She texted me afterward, saying that she wanted to keep our friendship, but that she'd been feeling attracted to me since we met. I read that message a lot. I still do sometimes.

Later when she snuggled up against me while watching TV, I took the opportunity to kiss her on the head. But she made it clear that she wanted our relationship to remain platonic, and from then on we took up our positions on different couches in my living room. We would hug once when she came over, and then once when she left, and that was the extent of our physical contact. But a month or two later she kissed me again, after a day at the beach, and after that I thought it was only a matter of time.

And yet, at the same time, I was certain it wasn't only a matter of time. I have always had a healthy fear of violating someone's consent mixed with an unhealthy inability to see myself as valued by others. So I never expected any interest from any woman, let alone one who identified as straight. Even as we acted more and more like a gay couple, running errands together—buying hiking boots at REI, making Target runs in her mom's Subaru—it was easy to just see us as being lesbians in every way except the most important way. It was a situation I was familiar with; Kelly and I always referred to ourselves as lesbians, despite both of us believing that I was a man all the way through our divorce.

Friends resented me for spending all my time with Genevieve, and were openly skeptical when I said that we weren't dating. We always insisted we were just friends. After all, when Genevieve stayed over we would spend the whole night lying rigidly motionless in bed, back-to-back, taking great care not to touch, which in retrospect wasn't the clear sign of noninterest that we'd taken it for. When we scheduled a trip to Sonoma for the two of us, everybody's skepticism jumped up a notch, including, if only subconsciously, our own. Nonetheless, when one of my closest friends asked me, the day before we left, to say whether I was in love with Genevieve, I said, "No." And I mostly believed it.

We had sex in Sonoma, of course. Everyone that knew us had been right, we were a couple. We'd just been the last people to find out. Two days later, she moved in for good. Later that week, as we were driving on I-880 as part of a failed attempt to adopt a rescue cat (we really lean into our gayness), Genevieve asked if I'd been in a serious relationship since my divorce and transition. I said that I'd had a few casual encounters, but no, not a real relationship.

"So really," I said, "I've never been anyone's girlfriend before."

"Are you my girlfriend?"

"I don't know, that's up to you. I hope I am."

"Yes, I think you are."

Girlfriend. Finally, a word that fit perfectly.

CHAPTER EIGHTEEN

How Did You Lose Your Virginity?

As far as I'm concerned, I was a virgin until I met Kelly, when I was twenty-five years old. Kelly was my first real kiss, let alone my first sexual partner. But I suppose there's an argument to be made that I'd lost my virginity a few years before: on a Saturday shortly before Christmas, in 2002.

It had snowed that night, and it was freezing. On the streets near my apartment, the snow had still retained a bit of that pristine beauty fresh snow provides, but here, down where Route 4 merges with I-75, Main Street looked even worse than usual, blackened, almost diseased, even where the businesses were still operational. I was driving to my credit union, a hideous 1970s-era building in beige brick, closed for the weekend, but with an ATM that didn't charge me a fee (something I was conscious of, having been raised by a man who considered being charged a fee to access his own money nothing short of fraudulent).

So I had driven my gigantic rear-wheel-drive Grand Marquis through the snow and ice to get some cash. I don't remember what purpose I intended to use it for, but the way I ended up spending it certainly came as a surprise.

As I sat in my car, steeling myself for another adventurous drive, I saw a woman approaching. She was walking from the adjoining parking lot, which belonged to a restaurant called, I believe, Benjamin's, though I think there were a couple words appended (Burgers and Wings?). I'd never been inside, despite living within a few blocks of it for most of my life, and thinking of it as a local landmark. It stood out on that strip of drab and abandoned buildings: a tall arched ceiling that always put me in mind of Vikings, somehow. Although . . . was I even sure I'd never been inside it? I had a childhood sense memory of potato wedges that I associated with that building, but no mental image of its interior. In any case, I understood it as a place that "we" didn't go, that served a different type of person than me and my family. It was for "those people."

Not that that was a term used in my house. My parents believed that part of what their faith required was to see everybody as a unique soul, loved and known by God, who asked us all to see them the same way. Nonetheless, they also knew who was "like us" and who wasn't. Certainly there was a racial component to it; Black people could be "like us," and many of them were, but the burden of proof was on them. The main criterion marking people out as "not like us," however, was poverty.

My parents had worked hard to remain "respectable," stable, financially self-sufficient, not taking welfare or needing to move in with family members. And I realize now that there were times when they had felt on the brink of failing at that goal. So people who had not met that standard, who had not remained "respectable," were dangerous, people who could suck them over the line, and doom them to life on the margins. And those dangerous people were the people who ate at Benjamin's, at least as I understood it. People who'd gone to public school, if you can imagine such a thing.

So the woman walking toward me was coming from a different world, one I'd been shielded from. She was around my age, maybe a few years older, several inches taller, light-years more confident. She was dressed nicely, but not appropriately for the weather. Her shoes were not designed to be wading through the slushy snow between the two parking lots. Her coat looked nice, if a bit cheap and showy, but in any case it was ornamental, not utilitarian. She didn't seem to have come from anywhere, or have anyone with her, and anyone walking through that snow in that outfit clearly didn't have an alternative way of getting around. And there was no question but that she was walking toward me. She seemed like someone who needed help, and I couldn't drive away without finding out.

She knocked on the passenger-side window, and I leaned over to crank it down a bit.

"Hi, can I sit in your car for a minute? I just need to warm up a little bit."

"Oh, um, sure, I guess?"

I leaned over again and unlocked the passenger door. She slid inside with a sigh of relief. I'd already started the engine, so the heat had been on for a bit. She sat there wordlessly, and I began to wonder how this encounter would end.

"So, um, are you waiting on anybody to pick you up?"

"No, I'm fine, I just needed to get warm. It's so cold today!"

"Yeah, yeah, it is."

"Hey, could you do me just one more favor? I'm really hungry, can you buy me something to eat?"

"Um . . ."

"You could take me over to _____," naming a restaurant whose name I've forgotten, more or less across the street, which sold chicken wings out of a building that was weirdly small for its lot; it gave the impression of having been part of a normal-sized building, the rest of which had been demolished. Which may have been true.

"Um, sure," I said. "I can do that."

I backed the car out of my space and began steering it carefully through the slush on Main Street. The woman seemed quite settled in, and I had no idea how to get her out of my car, to move her on her way and free myself up to do whatever I'd been planning to do that day. But, while she made me uncomfortable, she also had a sweetness about her, a sincerity and openness that made it impossible to imagine just kicking her out into the snow. But I needed her to go *somewhere*.

"So, where are you heading today?"

"Oh, I have to get back eventually, I guess. He gets mad when I'm gone too long. And I haven't seen my son in a while."

"Your son?"

Her face lit up. "Yeah, James. He's three years old. I kind of got into some trouble, and I couldn't keep him. I'm hoping I'll get to see him before Christmas."

I couldn't imagine what it would be like to be in her situation. Now I was happy to be helping her, at least a little bit.

We pulled up to the restaurant, and went inside, where there was a tiny area in front of the counter for people to order. No seats or anything, takeout only. She walked familiarly up to the counter and ordered some chicken, then added a couple more items to her order, glancing back at me each time. I told her I didn't want anything for myself, then walked up and paid for her meal.

We got back in the car and I started the engine up to run the heat while she ate. She had not been kidding about being hungry; she ate her food with a focus that made it obvious how much she needed it. When she was done, she asked me to take her on another errand, and I agreed. I don't remember what the errand was, but on my mind was the slowly dawning realization of what this woman did for a living, a realization that she soon confirmed.

"Thank you so much for this. You're really so sweet, it's so nice to meet a sweet guy."

"Um, no problem, you're welcome."

She put her hand on my knee. "So, do you want a date?"

I laughed nervously. "Yeah, I've been trying to figure out the answer to that." And I had been. I was a senior in college, as horny as any other testosterone-addled twenty-year-old, yet I'd never even made out with anyone, let alone had sex.

I'd considered finding a prostitute before, but I didn't have the slightest idea how one would go about it. As powerful as my lust was, it was nowhere near as powerful as my fear of embarrassment. I fantasized about going to a legal brothel in Nevada, where it was just like any other business, and I could be honest about what I wanted, and they could be honest about what they provided, and at what cost.

But here I was, with at least the initial part of the proceedings already handled before I even realized it. I hesitated, but there wasn't really a doubt what I'd decide.

"I guess . . . Sure, I guess I'd like that. Um, how does that work?"

"Do you live around here?"

"Yeah, not far."

"Why don't you take me back there? Show me your place?"

"Yeah, okay. And, um . . . How much . . . ?"

I don't recall the amount she quoted, maybe $100, but I had enough cash on me to cover it. We drove the few blocks to my apartment building. While my mind spun in circles, she held up the conversation, alternating between complimenting me, thanking me, and making oblique references to the hardships in her life.

I parked across the street and led her to the door and up the stairs to my third-floor apartment. I felt numb and disembodied. My legs were walking me home. Did I even want this? Of course

I did. Shouldn't I be excited? *Was* I excited? What was about to happen? Was I making a huge mistake?

My apartment, while less soul-crushing than my previous one, had minimal aesthetic appeal. It was a mess. Among other things, it is virtually certain that there was a pot lying on the floor somewhere, with the remnants of my last Kraft Mac & Cheese and a fork. Nonetheless, she was complimentary, telling me what she thought I'd want to hear, but still with a real sincerity underneath.

Decades later, when I started hanging out with more drug addicts, I could see how, relative to some places she'd probably been, my apartment *was* nice: spacious, reasonably homey, with the occasional appliance or item of furniture that had been bought new, and was still in good condition.

Well, I thought, *I guess this is it.* I still felt conflicted, but the horniness had started to predominate. *It's finally happening! I'm finally going to know what sex actually feels like!* But, like, what was the protocol in these situations? What were the steps between "accepted an offer from a prostitute" and "actually having sex with a prostitute"? I wasn't sure. I handed her the money and said something like "What now?" or "How do we . . . ?" or possibly just "Uh . . ."

"You want to see my tits?"

"Yes!" That was an easy one. I could imagine few situations in which I would say "No" to *that* question. She sat me down on my couch and took off the sweatshirt she'd been wearing under her coat. She wasn't wearing a bra, and there they were, some real live tits, maybe a C-cup. Because of all the porn I'd watched, my first reaction was that they seemed imperfect, but by the standards of actual human tits, they were quite nice. She bent over me on the couch.

"Would you suck on them for me?"

I expressed my assent, and she held my head and guided it onto a nipple, which I quite willingly took into my mouth. I don't know what I'd been expecting, but somehow it felt surprising. It felt bigger in my mouth than I'd thought it would, and it had the rubbery feel of . . . Well, it felt like a nipple, is what it felt like. But it was new to me. And while I was certainly getting hard, it was also unexpectedly nonsexual, in a way.

It felt exactly how I'd fantasized that it would feel, and I found myself disappointed. Apparently I had secretly been expecting that there would be some unanticipated sensation, something I'd been missing out on. Because if it was just going to be how I imagined it, then what was the point? I could imagine it whenever I wanted, and be by myself, not have to risk the horrors of trying to interact with other people. But on the other hand, it was still my fantasy taking place, and I was still at an age where I could get an erection in almost any circumstance, let alone when I had a girl's tits in my face for the first time. She was sitting on my lap and could feel it.

"Do you want me to suck on it some?"

"Sure, yeah. Sure."

"I just gotta put a condom on you."

This was good news. I would have been uncomfortable without a condom, as I vaguely understood all genitalia to be swarming with a variety of diseases with embarrassing names that would mark you as an outcast, whether or not they had any other lingering effects. I was all about safe sex, though it had been merely theoretical to date.

Condom applied, she began, you know, sucking my dick. And it . . . didn't really feel like anything. The condom was thick, and more important, I was so bewildered and anxious that I was kind of disassociated from my body. It was just what happened to me, when I was in a situation where I felt that other people were expecting or desiring me to behave in a certain way, but I didn't know what that way was. I'd be paralyzed, unable to act until I knew what was being expected of me, what script I was supposed to follow. My body would no longer seem connected to me, just a shell, an exoskeleton whose motions I could control only with conscious effort, and even then, not with any precision.

So there I was, alone in my apartment with someone who I couldn't really read, whose sense of "right"-ness I couldn't predict. Everything in my upbringing had been more or less designed to prevent me from learning how someone like her saw the world, from understanding what expectations she might have, from knowing how she might act in a situation like this one. And as horny as I was, the paralysis was stronger. I always felt it mostly as a difficulty in moving my jaw, in unclenching my teeth. A struggle with doing what's physically required to speak.

She came up for air, and tilted her head back to see how I was doing.

"Your dick tastes so good, baby," she said. No sincerity there, or much of anything else. Now that we'd gotten into it, she was just going through the lines of her script. "You want me to get on top of you?"

I wasn't great at knowing what I wanted at the best of times, which this was not. I forced my mouth open to say, "Yes," since I had to say *something*. She got back up on my lap, without taking off the purple snow pants she'd been wearing this whole time, and began humping me. This part of her script read "Make sex noises," so that's what she did, as she held my face to her tits and rocked up and down. If the blowjob hadn't felt like much, this truly felt like nothing. All the erotic feeling was draining out of my body, and I knew I was about to lose my erection, and however uncertain I was about the protocol, I did know that losing my hard-on was *not* what was expected of me. So I took my cue from her, and made some sex noises of my own.

"Unh, uh, yes, uh-huh, yes!"

She coasted to a stop, looked down between her knees to see my limp dick.

"Did you cum, baby?"

Shit. I also wasn't supposed to cum too soon. Had I screwed up anyway? Just get it over with!

"Yeah, yeah, that was really great!" I dug up a smile from somewhere, and put it on my face.

"Oh, yeah, it was great for me, too, I came a bunch of times." Still running her script as she put her shirt back on. She asked if she could use my bathroom, I said that was fine, and she went in, but came back out a few seconds later.

"Hey, so . . . Would you mind if I pooped in there, too?" She was off script.

Startled, I said of course she could, and she went back into my bathroom as I sat in a daze, trying to figure out what I could have done differently, what I wanted to do now. Could I still try and have sex? God knows I still wanted to, but would any further efforts be just as awkward and fruitless? I'd reached no conclusions when she emerged.

"Thanks so much. I usually don't do that in people's homes, but you're so nice, I really feel comfortable here." We were back to that sweet sincerity I'd noticed in her from the start. I was touched. I told her I was glad, and I truly was. I suddenly felt that all my discomfort had been worth it, if it had given some comfort to this woman, who clearly didn't have much of it in her life. Her next question surprised me.

"Can I borrow some underwear? Mine's kind of a mess, after I came on you like a million times." It was a strange thing to ask, especially given that that last part was clearly untrue, but I hadn't said no to her yet and I wasn't about to start. But what would she do with my underwear? It was for boys!

"Well, sure, but you know, it's just boxer briefs . . ."

"Oh, that's fine, I like those." I went into my bedroom and dug some out of a pile of laundry that I was pretty sure was clean. As she changed into them, I saw a brief glimpse of pubic hair, though I tried not to be too obvious about looking at it, out of an incongruous sense of propriety.

"You've been so nice. I really appreciate it. Can you give me another ride? I need to get to Matt's, but I'd really like to see you again. Would you like that?"

"Well, sure, yeah, I'd definitely like that." My horniness was reasserting itself, which she seemed to notice.

"Tell you what, you take me to Matt's and then wait in the car. I'll just be a few minutes, then we can come back here, and you can make me cum again with that big dick?" My dick was fine, but nothing special.

"Okay, that sounds good."

We went back to my car and drove a couple miles north, and she chatted a bit more about her life. I don't really remember much about her story. It was all so foreign to me that I couldn't piece together a coherent narrative. All I remember is the way she talked about her son, the anguish and longing that was barely under control, the deep guilt she felt at being separated from him, and about the life she'd found herself in. I didn't say much, because what was there to say? If she couldn't see a way out, I certainly wasn't going to find one for her.

She directed me to a house in a neighborhood I knew only by reputation. Back when I'd been a delivery driver for Domino's, our territory stopped a bit south of it; the franchise to our north always tried to talk our drivers into picking up some hours there, but we rarely accepted them.

Most of my coworkers had been held up at least once, and they didn't care to repeat the experience. But it didn't really look that bad. The number of vacant houses, and the general upkeep of the occupied ones, wasn't much different from where I lived. But there was a general air of wariness, a lot more visible security measures, barred windows and the like.

She got out and I waited in the car. I was psyching myself up; *this* time, I thought, this time I'm really going to do it, cash in my V-card, find out what pussy feels like after years of speculation. I'd just been off guard the first time, this time I'd know how to ask for what I wanted. (I'd have to stop at that ATM again, but I wasn't as broke as usual, I could manage it.) But she didn't come, and didn't come, and I was growing more and more uneasy as the minutes stretched on. After twenty or thirty minutes I was on edge, all horniness replaced by fear.

I didn't want to leave her there; she was clearly with her abusive pimp, right? That's how this worked? And if all I could do for her was give her a little time away from him, a little bit of safety, I wanted to do it, she deserved so much better than life had given her. Yet I was in a place I clearly didn't belong, and that I had been trained to avoid. And what was I going to do? Go knock on that door and say, "Hi, Mr. Pimp? Could I have your woman, please?"

My dilemma ended when a cop car pulled up beside me. I pretended not to see it, but they hit their siren for a second, and gestured for me to roll down my window.

"You've been here awhile, kid. What are you doing? You know anybody around here?"

"No, uh, uh, no, I was, look, I'm sorry, I was just leaving, I'll go home right now!"

I feel like this rendition understates my level of incoherence in this interaction, but in any case, it was what they wanted from me. I pulled out and drove home, heart racing the whole way. I'd blundered into the criminal underworld, and everything I'd picked up from pop culture led me to believe it wouldn't end well. I'd promised her I'd wait for her, and I hadn't. What would she do when she came out and I wasn't there? Would she be in trouble with her pimp? Would *I* be in trouble with her pimp?

My imagination was running wild by the time I got back to my apartment, filled with vague but lurid fantasies about a vengeful pimp coming to pay me back in unspecified ways for an unspecified harm. Plus, I had abandoned her! I'd been the white knight that was saving her, however temporarily, but now I was just a coward.

As I sat in my apartment, going over and over everything that happened that afternoon, the fear became more and more predominant. I was under no illusions about my ability to defend myself from . . . really, *anyone* who wanted to hurt me, let alone Vengeful Pimp, this overwrought fantasy I had conjured. She knew where I lived, so if Vengeful Pimp wanted to know, he would know, she wouldn't hide that from him, and I wouldn't expect her to, or even want her to. Whatever drops of comfort I'd given her weren't worth the sort of punishment I imagined she could receive for hiding information from Vengeful Pimp.

And I couldn't tell anyone, *anyone*, about what was happening. I'd paid for sex, I'd broken the law, and more important, I'd broken caste. Paying for sex wasn't something people "like us" did, at least not in such a low-class, tawdry way. Anyone I told not only wouldn't help me, but would disown me, cast me out of society. At least, that's what I was meant to believe.

There was nothing I could do, no way of avoiding the punishment coming to me: punishment for being horny, punishment for associating with "the wrong people," punishment for abandoning her, punishment for being awkward, for not going to church, for getting a D in my statistics class before I switched majors, punishment for what I'd been waiting my whole life to get punished for. I just had to sit and wait for it.

I did the only thing I could think of to protect myself: I shoved my couch in front of the door to keep anyone from kicking it in, as I presumed they would. It was a sofa bed, with a heavy metal frame, and would at least take some effort to get past. On that mid-December Saturday, it was already getting dark. I sat on my couch and tried to read a book for an hour or two, but mostly stared into space while spinning out various horrific fantasies. Eventually it was late enough that I could justify getting in bed.

As you may have guessed, I was not murdered that night. I was still on edge the next day, staying inside with the couch still guarding the door. There wasn't a person in the world that I'd have been glad to see that day. I just wanted to be alone and process my shame and guilt and fear and excitement and pride and lust and regret and confusion and so on. But having somehow gotten a night's sleep, I was at least starting to feel a little calmer.

It is virtually certain that I'd eaten a fresh pot of Kraft Mac & Cheese by the time somebody knocked on my door. I know that I was feeling marginally more calm in that moment, and at that time in my life, I associate calmness with KM&C. But when I heard that knock, everything flooded back. I ignored it at first, pretended I wasn't home. Then the knock came again, and I felt I had to answer. Even if it was Vengeful Pimp, I couldn't be rude.

"He-hello? Who is it?"

"Hi, Tom? It's just me." It was her. And while part of me continued spinning the fantasy that she was there with Vengeful Pimp, who was making her talk to lull my suspicions, I could tell that wasn't the case. There was no stress or worry in her voice; even if she wasn't alone, she wasn't with anyone she was scared by.

Embarrassment replaced my anxiety. What had I been thinking? What kind of nonsense fantasy had I come up with? Why did I think anyone would be coming after me? I hadn't *done* anything, after all. The couch was still blocking the door, and it wasn't like I could move it out of the way with any subtlety.

"Oh, uh . . . Oh. Hi!"

"I came out to see you yesterday, but you weren't there. What happened? Can I come in?"

"Well, um . . . Yeah, I mean . . . Hold on, I'll be right there." I shoved the couch out of the way enough to open the door, as quickly and quietly as possible, which turned out to be neither quick nor quiet. *God*, what is *wrong* with me, she's going to think I'm an *idiot*! I opened the door partway, and I peered out to confirm that Vengeful Pimp wasn't lurking in the hallway. He wasn't, of course, so I added that drop of embarrassment to the sea I was already drowning in, and slid out. She seemed a bit confused.

"Hi, baby, what's going on? Are you okay?"

"Yeah, I just . . ."

She looked in the doorway and saw the couch still partially blocking it, and I saw her face as she realized why it was there. Her face softened, and she looked at me with an expression of sympathy, and forgiveness, that remains my clearest visual memory of the whole experience.

"Oh, baby, you were afraid, weren't you?"

This was unexpected. She felt sorry for me! She understood my feelings, and cared about them! I did my best to hold back my tears.

"I just . . . You were gone a long time, and then this cop car pulled up, and they asked me what I was doing, so I said I was leaving, and then I was afraid to go back, and . . ."

"Of course you were! I'm sorry I took longer than I thought, and I'm sorry you were afraid."

And she reached out, pulled me close, and gave me a hug. I couldn't remember the last time I'd gotten a hug, a real hug. A hug that was an offer of comfort, from someone who saw my pain and wanted to ease it if they could. I hated hugs, associating them with family obligation and discomfort. But this hug was the real deal. There was nothing calculated about it, nothing patronizing, nothing to do with her job. She just reached out, one person to another, and I sank into her, and began to cry.

"I know, baby, I know."

A pause.

"So, do you want me to come in?"

"Well, I . . ." Despite her sympathy, despite how much I felt I cared about her, despite the part of me that still wanted to find a way to have sex with her, the twenty-four hours since I'd seen her walking toward me across the Benjamin's parking lot had been nonstop anxiety and confusion. Whatever else I wanted was drowned out by the overriding urge to be done, to end this strange little chapter of my life and return to my own world, where there was college and desk jobs and nobody was violent or talked about sex.

"It's just that . . . I don't know, I kind of have some stuff to do."

Her face fell. She accepted the situation and began moving on. I don't know if she was ready to be done with this chapter or not, but she knew it didn't matter if she was ready.

"Well, okay, maybe I'll see you around, if you want another date sometime."

"Yeah, sure . . . Yeah."

"Okay, bye, sweetie."

And without hesitation or a backward glance, she walked down the hall and back out of my life. I've never stopped thinking about her. About her life. About whether she was able to get her son back. About whether she thought of me, and associated me with Christmas, the way I did with her. I hoped that her life turned out okay; it's hard for me to imagine a great ending to her story, but I barely knew her story, so who am I to say?

Anyway, that was the only sexual encounter I had before Kelly. After that experience, was I still a virgin? That's open to debate. But I know I'll always be grateful to her, and glad that I happened to be in that parking lot when she needed a place to be warm.

CHAPTER NINETEEN

What Is Mental Health?

When I was nine years old, I woke up one morning and my mom wasn't there. I asked my dad where she was, and he sighed, then said, as if he were just remembering, "Well, I came home last night, and she was really drunk, so I had to take her to the hospital. She'll be okay. She just needs to stay there for a couple weeks while she learns how to not drink."

"Oh, okay," I said. Something along those lines, anyway. But I didn't mean it. What I really thought[1] was:

Bullshit.

I was immediately sure that I knew the real story. It seemed so obvious to me that Dad had never liked Mom, and now he'd taken advantage of an innocent mistake to pack her away somewhere where she wouldn't annoy him. He didn't "have" to take her to the hospital; he just didn't want her in the house. I was willing to concede that she might have been drunk that night, because otherwise she wouldn't have gone, and the hospital wouldn't have accepted her. But if she was an alcoholic, I would know. I'd never

---

1 In whatever language my nine-year-old self would have used.

seen her drunk, so I didn't think she had a problem. And I got all As in school. Case closed.

As far as I know, I *hadn't* ever seen her drunk, although of course that might have had something to do with being, you know, nine, and maybe not the best judge of adults' levels of intoxication. But at the time, I was pretty sure I *was* the best judge, at least when it came to my mom.[2] Like, she was *obviously* the greatest mom in the entire world: the most beautiful, the kindest, the smartest, the most fun. I was so lucky to have her! And yet, nobody else seemed to really share my opinion. I felt like my dad was mad at her all the time, and other people got mad at her sometimes, too: aunts, uncles, her friends, Grandma,[3] lots of people.

They never seemed to be able to understand why she was the way she was, how hard life was for her, all the obstacles she had to deal with, not least of which was all of these people in her life who kept getting mad at her. She probably told them that she wasn't that drunk and they just didn't believe her. Nobody ever believed her. So, obviously, my dad was just sick of her being around, and wanted to put her in this place where he didn't have to think about her for a while. He had waited until there were no witnesses, and then he played a trick on her.[4]

And now she was gone. The next time I saw her was in the

---

2   But I also thought I was the best judge of everything, tbh.

3   My mom's mom, about whom I now suspect my mother could write her own essay, on how daughters are affected by their mother's mental health.

4   I don't *think* I ever suspected my dad of having gotten her drunk intentionally, but I wouldn't put it past me back then. And I also want to use this footnote to clarify, in case it's needed, that my dad was a basically decent guy who handled this whole thing pretty well, under the circumstances.

psych ward at the hospital, a place that, while it had a different feel
to the hospital I went to every year when I had asthma attacks, was
still one of those places people made you go, like school, or church.
We met her in a room with toys in it.[5] My mom came in wearing
a hospital robe. Why? She wasn't sick. We were all wearing regular
clothes. Why were they shaming her like this? The hospital lady
asked us questions in that singsongy way, where you can tell they
ask this same question to a bunch of different children every day.
At one point she asked me, "Can you draw a picture of how it feels
when your mom drinks?"

An assignment! I knew what to do with that. I took the available crayons, of varying length and sharpness, and after a little
thought I drew a picture that I was pretty sure was what they
were looking for. She seemed to like it. *Sweet, I get an A in family
therapy!*[6] Then she said, "Do you want to talk about this picture?"

I didn't know there would be follow-up questions.

"Yeah, I don't know. I've never actually seen her drink? But
this is what I *imagine* it would be like if she ever did drink."

I said it in a bitchy way, too. Sarcasm had long been almost my
only tool of rebellion. I could never be seen as rebellious, however
I felt. I had to follow the rules. And so if I was going to resist, I had
to do so within the rules. I had to find loopholes. All rules have
loopholes. A ton of them can be exploited by sarcasm, by insin-
cerely saying what the other person wants you to say, and daring

---

5   You know the kind I mean. Waiting room toys. Toys that belong to adults
who work in places that children are required to be in. Toys treated with all the
care that children traditionally use with other people's possessions. The tragedy
of the commons personified. Disheveled, I would call them.

6   I didn't actually expect to get a literal grade. But as far as I could tell, kids
were graded on *everything*.

them to call you out on it. The family therapy lady moved on to my brother after that.

Every day my mom was in the hospital, someone in our parish would come around and bring us dinners that they'd cooked. This was kind of them, really, genuinely commendable behavior, one of those things that reminds me of what can be so valuable about having a church community. It also made me furious. I was a picky eater as a child, and my mom was a good cook.[7] So now dinner was being made by people who weren't as good and also didn't know which vegetables I was unwilling to negotiate on. And even though it was from a rotation of families, the meals seemed monotonous to me, and probably were. It was emergency dinners made by housewives in Ohio in 1988. How many chicken potpies can a person eat?

And it was humiliating! As if we needed food brought to us. As if we couldn't take care of ourselves, couldn't feed ourselves without Mom there. Actually, she hadn't been there lots of times! She would get a headache, one of those headaches where she had to stay in bed for a few days, and nobody had brought us dinners then! But now, just because my dad had made up this bullshit reason to get her out of the house, everybody felt sorry for him, even though this whole scheme was his idea! Feeling sorry for us when we didn't ask them to. We didn't need it, we had this! Everything was under control!

I mean, it's pretty funny, in retrospect. What did I think I was going to do? Cook dinner? I was nine! But I was confident. And I

---

7   I mean, obviously most people think their mom's cooking is the best. But certainly she was adventurous, by prevailing standards. She cooked a lot of what people called "world cuisine," which in the '80s essentially meant food that contained grains other than wheat and corn, and colors other than beige. I regret how much of her effort was wasted on me.

didn't notice any connection between that and the lack of confidence I felt in my parents, never more so than during the Month of Chicken Potpie. They had lied to me! My dad had lied and said my mom was an alcoholic, and now my mom was telling the same lie, saying that she needed to be in the hospital, just so my dad wouldn't look bad.

I would have to handle everything myself, just like always. From then on, I considered myself the real head of the household, our public face and internal decision-maker, whether or not anybody else realized it. All this had proved was that I didn't need any help from anybody. And that every story I heard from grown-ups had better be fact-checked before I trusted it too far.

My mother has been sober ever since, but for decades, I stood by my story: my mom had never been an alcoholic, she just liked AA meetings because she could talk about herself and people had to listen. I know that's pretty harsh, but I judged my mother pretty harshly in those years. I denied that any of the failings I saw in her could have affected my life in any way. Because I'd been the real adult the whole time, I'd always found what I needed for myself. The Month of Chicken Potpie made me doubt my mother's victimhood. Now, I didn't resent the grown-ups around me for not believing her. I resented them for believing her too much.

I started realizing how many of her "headaches" corresponded with family gatherings she didn't want to attend, that she was just bailing on them. And I would see my dad make excuses: "Betty's not feeling well." I hated to see how embarrassed he was when he said it, and I hated even more how embarrassed it made me, when I was obligated to go along with a story that I didn't believe, to look like a rube who didn't know my mom was faking it. From ages ten to eighteen, I remember countless occasions when we were heading out to something, or even just about to sit down to dinner, and she'd say, "Oh, I'm not feeling well. My head hurts.[8] I'm going to go lie down."

Each time I would think, *Well, maybe it's true, maybe you're not feeling well, but guess what, you haven't been feeling well in years. So apparently, this is just how you are now. And you can't act like this is something different. You can't pretend that this is just how you feel "right now," because it will always be "right now." You need to figure out how to live with it and show up. Be part of the family. Help us get through this miserable life we all share. Don't abandon us again.*

And then, a decade or so after moving out of the house, I married Kelly. And on multiple occasions, one of my relatives, usually one who had married into the family, took Kelly aside and said, "Here's what you need to know about Betty." They told Kelly stories from my childhood that I'd never heard, stories about her drinking, stories that made her sound like, well . . . an alcoholic.

Which was another betrayal. *Why did nobody ever tell* me *about Betty?* I wanted to yell at my family: *You all saw me being raised in chaos. Why didn't anybody ever talk to me about it? Why didn't anybody ever validate my feelings of not liking her? Not liking my own mother? Do you think I enjoyed that? I thought I was a monster for feeling like that, and you never thought to tell me you felt the same way?* That hurt.

A year or two after meeting Kelly, I was cast in a play by a friend of hers. Through the theater department she was attending, students could put on their own plays, and get time in one of the performance spaces to put it up, and one of them invited me to play a role in their production of an absurdist play. As a community theater veteran, to be considered worthy of acting alongside

---

8   Or her stomach, or a cold she'd had at some point in the last month was "still acting up," or whatever felt the least implausible to her in the moment.

these people who took theater so much more seriously than I did was a huge ego boost. And I think I'm being honest when I say: I did a great job! I absolutely felt like I belonged on that stage. It remains one of my fondest theater memories.

But the other memory I have is after the play, when the house-lights went up and I looked out into the audience and saw my dad and brother there. But not her. My dad told me my mom's stomach wasn't feeling well, with that same embarrassed grin I'd seen a thousand times before, and my heart sank. I was so upset at myself. *This whole empty-chair-in-the-audience thing is such a cliché. What is this,* Degrassi Junior High? *I'm a grown-up. I don't even like my mom!* But I couldn't reason myself out of the pain I was feeling.

So what did this whole experience teach me? I don't know. I'm still figuring that out. But I've come a long way. And the main reason I've come so far is that, a few weeks after that night when I stood with my father and brother, fighting back tears in the white institutional hallway of the Creative Arts Center (so much like all those other hallways my mom had disappointed me in), I walked into the office of a stranger, and said, "Well, I'm not really sure about this. I don't think I really need therapy. But a few weeks ago, something happened with my mom . . ."

I've made a lot of good decisions in my life. That one might have been the best. If you've ever considered seeking therapy, please do it. It might take you a few tries, but it can transform your life in ways you aren't even imagining. You might not think you need it, but then, I used to think that a nine-year-old could care for a family of four. Stop lying to yourself, and start lying to a professional. And eventually, you might be able to start telling the truth.

CHAPTER TWENTY

What Is a Home?

Where are you from?

It's one of the classic small-talk questions, and I have part of my answer down pat:

"I live in Oakland, by Lake Merritt, near Fairyland. I was born and raised in Dayton, Ohio, but I've been in Oakland since 2009, and I don't ever plan to leave. I fully intend to die in Oakland. It's home." Then, if I'm asked about Dayton specifically, my usual line is "It's a great place to be *from*," with both the insult and the compliment intended. I do love Dayton, and I don't regret my time there. But I'm glad I don't live there anymore.

People sometimes ask *why* Oakland is home to me[1] while Dayton is not, and I always find it hard to explain. If you want to know why Oakland is great, there's no substitute for just moving there.[2] It worked for me. While I never seem to be able to explain why

---

1 People who don't live in Oakland, usually.

2 Also: Watch *Blindspotting*, both the movie and the TV show. I know I'm skewed, but I think they are both incredible works of art, whether you know Oakland or you don't. But knowing Oakland helps.

Oakland is the greatest city in the world, I can at least try to capture some of its appeal.

Apart from early childhood, the first time I felt at home was my senior year of high school. It didn't have anything to do with where I lived,[3] it had to do with who I hung out with. I had friends who I was close to. I knew my place in the social scene. I had a driver's license. I was having new experiences all the time. It was the best time of my life to date, and I knew it. I would say to myself: *This is it. This is the good life; this is my peak. College won't be like this, and probably not adult life, either. I'm not going to get this back and I need to enjoy it now.*

Here's a random example, which probably isn't convincing: One day my friends Ben and Beth picked me up from a *Twelfth Night* rehearsal. They told me that we were going to see *Pulp Fiction* again, but then we got there and they said, "No, actually, we're watching *Sid and Nancy!*" I hadn't wanted to see the movie, and they knew that. But they wanted to see it and they wanted me to come, so they dragged me along, knowing that I wanted to be tricked into spending time with them, even if I didn't know it. I didn't enjoy the movie that much, but nonetheless, it's one of my happiest memories.

I loved that I had a role in their lives and a place in our little friend group, that I was the one who they wanted to drag to the movie they wanted to see. They didn't drag me to the movie because they were bullies, and they didn't drag me to the movie because it was a religious obligation. They dragged me to the movie because they thought that I would like it, and they just wanted me

---

3 I slept in the top bunk of a bunk bed, crammed in a closet-sized bedroom only slightly bigger than the bed itself, in a decaying house in a decaying neighborhood. During my childhood the neighborhood decided to close off a bunch of streets to cut itself off from another part of town (*cough*Black people*cough*).

to be there with them. It was the most accepted I had felt in my life, and I knew it was about to disappear.

And it did. Beth went off to art school in New York, because she's one of the coolest goddamn human beings I've ever met. Ben went off to school in . . . Well, he went to Toledo, but at least he went somewhere. I stayed in Dayton. If I'd worked harder in school, or if my family had more money, I'd have gone somewhere else. But I hadn't, and they didn't, and my dad worked at UD, so I got free tuition, so that's where I was. The University of Dayton, despite its many fine qualities,[4] was perhaps the least homelike place I've ever been. It was where rich Catholics who couldn't get into Georgetown or Notre Dame went. Frats ruled the campus. I hated it there. I had to live in student housing the first two years, and without getting into too much detail, I swore a solemn oath to myself after sophomore year that I would never live with roommates ever again. Those dorms were never home, they were just beds assigned to me in a nightmarish summer camp that lasted for eight months.[5]

The first place I lived on my own was a furnished one-bedroom at the top of a hill near campus, and while I hoped it would feel like home, it never did. The place was big and empty except for random pieces of other people's furniture. The landlords lived across the street with their large family, several of whom always seemed to be hanging around the office whenever I went in to drop off my rent, staring wordlessly at me until I left. I was afraid of them. Though, to be fair, I've been afraid of pretty much every landlord I've ever had. I'm sort of like a cat—I'm territorial. I need my space that is *my* space, and that I can feel ownership

---

4   I assume.
5   How long is a college school year? I really don't remember.

of and safe in. Landlords remind me that where I'm living is not really my space and can be taken away from me if I misbehave, or if they just decide they'd rather I wasn't there.

It was pretty lonely. One day a fly got in. I spent a few days trying to kill it with a rolled-up newspaper, and it kept evading me. It got to where I almost respected the fly for being so good at surviving. Then I went to the store and got an actual flyswatter. It turns out there's a reason there's a specific design for flyswatters. The first time I took a swing at the fly with it, I killed it. And I was devastated. We'd had a whole relationship going, that fly and me! That poor fly had never had a chance. So now it was just me again, sitting on a couch that didn't belong to me, that had never belonged to anyone, watching the room darken and feeling a creeping sense of dread.

Meanwhile, I kept getting glimpses of a world where I could feel at home. I visited Beth at Cooper Union, and while I was there, I went with her to a party with a bunch of art students. We got drunk and ridiculous and happy and had absurd conversations that I knew I would think about for a long time afterward. A woman hauled out her acoustic guitar and sang "Both Hands" by Ani DiFranco. The rest of the room rolled their eyes out of their eye sockets,[6] but I was enthralled. I'd never seen that happen at a UD party.

This is where I should be, I thought, looking around that room. These were people with whom I could discuss books and Taoism, people who would drag me to movies because they wanted me around. Yet I never thought I'd end up there. I kept moving around Dayton—the ghastly studio apartment where I hit rock

---

6   It seemed clear that this was not the first time she had claimed to be responding to a request for a song that nobody had made, or ever would.

bottom; the seedy apartment in a nice-ish neighborhood by the art institute where my neighbor broke into my apartment to steal change off my dresser, and from which I secretly moved out without notifying the landlord, for reasons I can't explain; the bland, beige two-bedroom in an apartment complex in West Carrollton.[7]

Eventually Kelly and I moved into a two-bedroom apartment that, while it didn't have much personality, was as "home" as I'd ever expected to be in Ohio, or at least until we bought our own decaying house to one day die in in a decaying part of the city, as all our relatives had done. Kelly was the reason I moved out of Ohio at all. Meeting her gave me the life I'd been waiting for: smoking weed and hanging out with weird theater kids. She opened me up to so many things, but maybe the best of them was the West Coast.

We first went out there to visit some friends of Kelly's, one of whom was in grad school at Berkeley. It happened to be Pride that weekend, so we had to go check it out. Before we left we ate some homemade weed brownies that turned out to be *extremely* strong, but once that calmed down a bit, it was quite an experience.[8] We marveled at the diversity; in Ohio "diversity" meant that there were straight white people and straight Black people in the same place. But here, every different type of person you could think of was there, having a good time and being who they were. That struck both of us as our kind of environment, despite our being, as far as we were aware, a heterosexual couple.

So that was one of the first major appeals. And then we also

---

7 The apartment complex was called Indian Lookout, and my home was on Shawnee Drive. Sigh.

8 I'd say more about what happened while we were stoned out of our minds, but it's one of my more reliable party anecdotes, and I don't want to burn it for this book.

went and saw a random play that we happened to see a poster for, and it was *so good*. It was a John Guare play from the late '70s called *Bosoms and Neglect*. There are two rich New Yorkers in the waiting room for their psychiatrist, who's late. And it's basically them talking about all of their therapy issues, nothing too memorable, but the performances were fantastic. That we could see something that good, just on a whim, thrilled both of us.

Plus, of course, California has great weather, and the landscape is stunningly gorgeous. Every week I'm out here, I see something that would've been the most beautiful thing I'd ever seen back in Ohio. People in California leave their windows open all the time. Constantly being in the fresh air, indoors or out, was something I'd never considered in Ohio, where you needed either heat or air-conditioning for fifty-one weeks out of the year. At the farmers market there's always incredible produce, peaches and plums and strawberries. For a month or two each summer this one farm sells blueberries that are so amazing they've ruined all other blueberries for me.[9]

Still, I had never thought that I would leave the Dayton area. Leaving seemed to be for other people. In my family, no one moved away. I had the feeling that living somewhere beautiful wasn't my fate. The fate of my family was to be in Southwest Ohio, and that was that.

Plus, moving away was scary. I knew where everything was in Dayton. I knew how to get from anywhere to anywhere, and I didn't get lost, even though I didn't have a smartphone. I knew people there. Not a ton of them, really. Most of my friends were Kelly's friends, and she had just graduated college, so a lot of them had moved to various places around the country, but there were

---

9   Not that I'm complaining.

still a few we hung out with. So the thought of going somewhere so far away was definitely intimidating.

But I also did what Kelly said, and Kelly said we were moving to California. The whole time I kept not believing that we were doing it, even as we found an apartment and hired movers, packed up all our stuff. But as it sank in that we were actually moving, I realized that I didn't really think I was ever coming back.

All of our relatives on both sides believed we would return, and pretty quickly at that.

"You'll miss the seasons!" they said,[10] or "It's so expensive out there!" or "Isn't Oakland really dangerous?"

But what I heard was: "Our kind of people don't go to California. You won't like it there. They'll be weird and different, and you'll want to come home to old familiar Ohio."

I doubted it.

So we drove across the country to our new Oakland home. I remember the night that we arrived,[11] the first thing we did was go find a Chipotle[12] because we were craving burritos and familiarity. For some reason, Google Maps took us to an abandoned military base on Alameda where there was nothing at all, let alone a Chipotle. And so that was our introduction to Oakland: driving past a bunch of boarded-up barracks in search of a fast casual restaurant,

---

10   I have never once missed the seasons, even a little. You know what's better than having seasons? Having the weather be nice all the time. I know some people really do prefer having seasons, but it's completely foreign to me. Why would I want to be *less* comfortable?

11   It was the night Oscar Grant got shot. Police helicopters were buzzing around us all night. We had no idea why.

12   Listen, leave the only home you've ever known, drive all the way from Ohio to California, arrive in a bare apartment where you'll be sleeping on an air mattress because all your furniture won't arrive for another day; then and only then will I allow you to criticize this decision.

speculating about the possible reason for the rioting that seemed to be going on.

A few days later, we went to a party a block from our new apartment with some of Kelly's friends' friends. It reminded me of those parties in New York with Beth. Only now, I wasn't a tourist, an interloper. I *did* belong at this party, with these people. I wasn't just a hick from Dayton anymore. I can't say what it was about them that I responded to specifically, and I'm not sure that I ever met any of them again after that night, but they had interesting conversations. I didn't have to be careful to not sound too smart or to say things that they didn't understand. Because if I did use a word or a concept that they weren't familiar with, that wasn't a problem for them. And in Ohio, it had always seemed like a problem. It felt like it was my lot to always get in trouble for just saying something that was on my mind or for knowing stuff that made other people feel dumb. But from that night on, I was in Oakland, I belonged there, and that was kind of that.

A day that I often use to try[13] to sum up why Oakland is home happened almost eight years later, in November of 2016. Four months after Kelly had announced she was leaving me, one month after she'd left, some unknowable amount of time between when I suspected I was trans and when I knew. Election Day. Mara and I went to a watch party at the New Parkway Theater[14] and watched the results come in with a bunch of our fellow white Oakland liberals. We went in super excited, ready to celebrate our first woman president. And then, well . . .

---

13   And fail.

14   A great institution, one of those things that could only have been made by the people who made it. In the alternate timeline where its founder, J. Moses Ceaser, gets hit by a bus, there isn't a lesser version of the New Parkway in Oakland. It just doesn't exist. Those are the kinds of places I love, that Oakland loves.

Once it became clear[15] what was going to happen, we left. Mara got an Uber back to her place in San Francisco, and I walked the ten minutes back to my apartment, feeling numb, exchanging a few robotic text messages with friends who felt the same. I passed multiple people openly weeping. Everybody was dealing with it differently. My way of dealing with it was to say to myself, "Welp, this seems like a great time to get drunk."

So I went to my local corner store, where I would go whenever I needed cigarettes, or Dark Horse Sauvignon Blanc, or when I was high and needed Munchies or Sour Patch Kids or Ben & Jerry's. It was a well-stocked Ethiopian grocery, always reconfiguring their property to try out different business ideas.[16] It had a few shelves full of Ethiopian[17] spice blends with no labels except the prices, making it clear that if you didn't know what they were, they probably weren't for you. It was one of my neighborhood's greatest resources, as far as I was concerned, an appropriate place for a wide variety of situations and moods, including the black hollow despair I was feeling then. That night, like most times I walked in, there were some people[18] hanging out near the door, chatting with the guys[19] behind the counter. The scene in that store didn't feel much different from any other time I'd been there. It was just another day to them.

And I had a new[20] realization: They'd all known already! Of course they had! They had known all along that this was the coun-

---

15   To us, at least. There was a faction of the crowd that remained in the theater, talking themselves into believing that things were still okay.

16   Remind me to tell you about the Icey Cream someday. Or the hookah lounge. Or the dry bar.

17   I assumed.

18   Non-white people, which will be relevant shortly.

19   Who also were not white, as I'm guessing you assumed already.

20   And to be clear, extremely belated.

try we lived in, a country that could choose to put a vulgar, sexist, racist idiot in charge, not despite but because of his failings, because their primary motivation was to humiliate and belittle everyone that they didn't like. A nation of bullies, of narcissists, of racists, and sexists. I was shell-shocked because I hadn't known that, but that was on me. The evidence had been clear that this was a plausible thing for America to do, and I just hadn't believed it. But the people hanging out, shooting the shit with the guys working that Tuesday shift behind the counter at the Ethiopian grocery, they'd seen it long ago. So why should they be so upset?

"Diversity" can seem like a vague virtue-signaling buzzword, a hazy concept of everyone just getting along. But on nights like that, I was so appreciative that I lived somewhere where I could be exposed to a different viewpoint than the white liberal one I'd shared with all those despairing or delusional people at the New Parkway. The benefits of diversity aren't always apparent, but when you need it, there's no substitute.

In a way, that night was my first experience of being in a minority class. I hadn't really come out to anybody as trans yet. I'd barely even come out to myself. I'd been living as a straight white dude my whole life, and I worked in tech. Sure, I was sympathetic to minority viewpoints, I was feminist.[21] But, even though I was still in a bit of denial about being trans, I knew that a small but real part of the Trump/GOP policy, which had just been endorsed by the country I lived in, was that Americans needed to be protected from, well, me. That was a new feeling. Never before had I had the experience of being targeted by politicians, targeted because many of my fellow citizens wish me ill, and base their vote at least in part

---

21   Whenever I meet a sincerely feminist man these days, I start to suspect they might actually be a trans woman. I hate that I do that, but I have at least occasionally been proven correct.

on whether or not it will hurt me. Of course, that's a feeling that the majority of the people[22] in the country have experienced for much of their lives, but I hadn't yet. But if you want to be around people who get it? Who know what that's like? Move to Oakland.

But don't go to Chipotle. Get a real burrito.

---

22 Everybody except straight, cis, Christian, white men, basically, along with people who are fooling themselves.

CHAPTER TWENTY-ONE

Did Your Parents Ever Give You "the Talk"?

Like anyone raised in a conservative Catholic household—and thus like essentially everyone I knew as a child—I'd been shielded from any knowledge related to sex. That this was the right way to raise children was unquestioned; indeed, the general attitude seemed to be that, in an ideal world, even adults would know as little about human sexuality as possible. And since I was attuned to what grown-ups expected from me, and was terrified of disappointing them, I did my best to avoid testing the boundaries of that prohibition. However, I also loved learning, in part because it gave me the parental approval I so desperately craved. And occasionally, that love of learning led me to ask questions that got me in trouble; or at least, caused awkward moments with grown-ups, which felt like being in trouble to me.

An early memory, from when I was maybe six years old: it was Sunday, right after mass, and we had crossed from the church over to the basement of the associated school next door, where we had donuts. We attended that coffee hour every week, provided I was well behaved in church, which I virtually always was.[1]

---

1   I was an "easy child," my mom always says, although I remember myself more as a "frightened child."

And really, even on days where I'd been restless, not content to sit quietly and read my Bible picture books, we usually got donuts anyway, since depriving me of donuts would mean that my parents couldn't have any, either. And who doesn't like donuts?

We would sit and eat our donuts with the rest of the donut-fancying parishioners, in a concrete basement room in the school building, with battered folding tables and chairs set out for the occasion. It was very familiar to me.[2] The yellow walls had posters promoting various religious, or at least vaguely positive, messages. Occasionally, whichever poster was the most tattered would get replaced with a new one, and the new entrant this week was a black-and-white picture of a child who looked sad, with the simple message "Stop Child Pornography." I saw it as we were looking for an open table, and then I asked what I believed to be an innocent question: "What does 'pornography' mean?"

I don't remember how my parents answered. All I can remember, vividly, is their discomfort. I was horrified. They were embarrassed! I'd said something wrong, something bad! I was in trouble! They quickly stumbled through some incoherent attempt at finding a child-appropriate way to define the term, and I nodded my head vigorously, making clear that I wouldn't have any inconvenient follow-up questions. They changed the subject.

There were a handful of similar incidents over the years, and I'd kept mental notes about which topics were not to be discussed, such as pornography, condoms, my penis and the way it got bigger sometimes, why people were embarrassed that our parish priest had gotten AIDS, what "gay" meant, what a "period" was, and on

---

2　For example, it was in this room, at around this same time, that I participated in the first-grade Martin Luther King Day pageant.

and on. And I worked hard to steer around those topics, to avoid that awful feeling I'd had in that school basement, the feeling of wrongness, of sin, of having hurt my parents somehow. But one summer day in 1992, a day came along that I'd been dreading:[3] my dad was going to give me the Talk.

I knew this rite happened to boys, that one day their father took them off somewhere and explained . . . You know, stuff? . . . *Grown-up* stuff? You know . . . *ess-ee-ex*? I'd seen it on multiple family sitcoms, and I didn't expect it to go any more smoothly for me than it had for, say, Cory Matthews, or Mark Taylor, or Brendan Lambert. I'd had a longer reprieve than I'd expected; the only reason it was happening now was that my dad felt uneasy that it had already happened for all the other kids in my class. Puberty wasn't like the spelling bee or Cursive Club. For once, I was coming in dead last out of my entire class.[4] And so, one day he announced that he'd made a decision: that Saturday, even though I didn't need it yet,[5] he would give me the Talk. Maybe it was still early for me, biologically speaking, but that wasn't a problem; even if my genitalia hadn't yet gone into production mode,[6] the Talk didn't include any practical demonstrations. There was no lab work. It was purely theoretical.

So the plan was, he'd drive me out to some park, we'd walk through the woods, and Dad would pass down the secret knowledge as the sunlight dappled the leaves. And like every other male-bonding plan my dad made, this one went south almost immediately. For starters, while I didn't actually resist it as much as

---

3   Though to be fair, that was true of most days when I was thirteen.

4   So, it was like penmanship.

5   "And never would," I wanted to say.

6   Computer science joke.

I had, say, clothes shopping, or Boy Scout camp, I nonetheless made it *quite* clear that my participation was strictly out of obligation. For another, my dad was just as flummoxed as I was by the conundrum of how to talk about . . . *ess-ee-ex* with one's kid. Because my dad was, before anything else, a rule-follower. And one of the most clearly established rules that he knew of, a rule that was written in the deep magic from the dawn of time, was "One does not discuss any orgasm-related topics with children." Yet there was another rule that said that knowledge about orgasm-related topics had to be passed down from father to son; indeed, that all the other orgasm-related rules[7] were only ever violated by sons whose fathers had failed to pass this knowledge on properly.

And it was raining. Of *course* it was raining.[8] It wouldn't have

---

7   And there were *so many.*

8   This wasn't the only day in my tween years when I correctly predicted rain well in advance; the other being the day of our seventh-grade class trip. On a given day at the end of the school year, every class had an outing: first through sixth graders went to a roller-skating rink, eighth graders went to Kings Island, a legitimately awesome theme park about thirty miles south, and seventh graders went to the Beach, a water park located at the same exit as Kings Island and positioned in the marketplace as the more affordable place to turn your children loose for a day in the summer. Our seventh-grade teacher, Miss Sippell, was new to the school (I think it might have been her first job out of college) and we all smelled weakness from day one. For example, one time J. D. Williams, the class clown, went and rooted around in her purse, just because she wasn't looking and nobody believed in her punishments. He saw that she had a little canister of pepper spray in there (our school wasn't in the *safest* of neighborhoods), and sprayed it into the air in the classroom, just to see what would happen. Anyway, the point is, we had known since we were in first grade that one of the rewards for not getting kicked out of school in the first through seventh grades was a free trip to the Beach. And we also knew that our free trip could *only* be redeemed on a specific date, and if the Beach was unavailable on the date due to thunderstorms (a not uncommon occurrence in Ohio in late May), we would be obligated to accept some substitute

been a father-son bonding experience if it hadn't been doomed from the outset, at least in our family. My dad still grimly dragged me out to the car. It had taken both of us this long to get our courage up. God knew when it would have happened again. But once we got out on the road, he admitted that we'd be skipping the walk.[9] We would just have the Talk right there, in the car, driving an aimless loop around exurban thoroughfares with names like Grange Hall Road, Colonel Glenn Highway, New Germany Tre-

---

excursion. So, on multiple occasions during that school year (more and more frequently as it neared its end), one of the bolder members of our class would ask Miss Sippell the question of what we would do if, on the day of our trip to the Beach, it was raining. And her response was always the same: "It won't rain." Of course, I wouldn't be telling you all this if not for the fact that, on the fateful day, it did rain, quite dramatically. And so the backup plan she was forced to make up on the fly turned out to be: take the class to McDonald's. Hang out there for a couple hours, even as they switched over from their breakfast menu to their everyday menu, resulting in at least one student (me) getting an order of french fries with no salt because the person who was supposed to salt them hadn't clocked in yet (I genuinely don't know how to reconcile the vividly remembered data points logged by my thirteen-year-old self in a way that makes sense to a grown-up, but one thing I am certain of is that I ate saltless french fries that day), while the teacher and the parents who hadn't been able to get out of chaperone duty came up with a new plan, which was: go back to our seventh-grade classroom, wheel in the media cart, and watch (1) *Father of the Bride* (the Steve Martin remake), followed by (2) *The Breakfast Club*. So the "treat" that we got (apart from McDonald's) would be the chance to sit in our everyday classroom and watch a couple fairly up-to-date movies on a combination TV/VCR that had been rolled in from the A/V room down the hall. And the *real* treat was that they wouldn't make us get permission from our parents to watch them. Which, sure, I would never have been able to watch *The Breakfast Club* at that age in any other circumstance, but it was small compensation compared to riding the Cliff without my parents there to tell me I was too small for it still.

9   Which was fine by me. I wasn't a huge fan of the outdoors. It was so dirty! There were bugs and worms and weird creepy fungal growths in it!

bein Road. Places like that, with nature safely beyond the shoulder and drainage ditch, and with few traffic lights to impose any unwanted rhythms.

The truth is, it wasn't raining *that* hard. We were both dressed for the rain, we still could have gone forward with the original, Mother Nature–approved plan. But my dad was no more eager for this experience than I was, and he clutched at the excuse the rain provided, to cut the whole event mercifully short. He was hardly going to get any pushback from me. He just drove aimlessly around, and began trying to Talk to me. The first thing he had to do was he had to somehow ascertain whether I had already figured out that penis-in-vagina was, like, how the whole thing worked. I had, but I was no more willing to say so directly than he was, so we spent a while skirting around the point as delicately and laboriously as a courting couple in a Jane Austen novel trying to gauge each other's interest in a threesome.

That accomplished, he moved on to what, to his credit, he clearly regarded as the primary point of the outing: Did I know that it was necessary to have consent from a girl before I had sex with her? He asked if I knew what "rape" meant. I did, but I struggled with how to phrase it. After all, in every context until then, the right answer would have been that I *didn't* know what it meant, but clearly that didn't apply here. Finally, I remembered a scene from the movie *Look Who's Talking*.[10] "Um . . . rape . . . um

---

10   If you're not familiar: the "who" that was talking was a baby, voiced by Bruce Willis, but only to the audience; in-movie characters such as the baby's mother (played by Kirstie Alley) did not hear their baby's sarcastic, surprisingly sophisticated wisecracking. That I found inspiration from this movie in a crucial moment may seem silly to you, but keep in mind that *Look Who's Talking* had recently been dubbed the "Favorite Movie" of America's children, according to the sophisticated analysis used to identify the recipients of the Nickelodeon Kids' Choice Awards.

... that's sort of ... when you have an affair with someone who doesn't want you to ... you know, do that," I offered.

I don't remember his response, but I will always remember the deep sigh that preceded it. My dad knew that I knew what he meant. And he also knew that I'd only phrased it in such an awkward way because that was what I had been taught to do. But it also wasn't a good enough answer, and so now he would have to call on even deeper magic, from before the dawn of time, and actually be clear and explicit about the nature of sexual intercourse, to make sure that his kid didn't hurt anyone.

Anyway, he pieced together some kind of response, clarifying that "an affair" was not really the mot juste in this context, and what specific acts were being referred to, and I nodded as hard as I could. *Oh, yeah, of course, I know, I mean, I just wasn't sure; I get it, though, I know what you mean, I know what it means.* Or whatever it was I said. And my dad made some sort of acknowledgment, and offered some sort of appreciation for the fact that I was already opposed to sexual assault even before the Talk. And that seemed to be about it.

It might have been at this point that he gave me a book. Or possibly just told me that he'd give it to me when we got home.[11] The book was called *Becoming a Man*, and it was written by a Roman Catholic monk, for just such an occasion. It was the Talk, in book form, guaranteed to neither (a) promote anything con-

---

11   I don't really remember the logistics. Where did he have the book stashed so that he could hand it to me? Did he give it to me right at the beginning of the Talk? I feel like the book must have either been (a) not in the car or (b) known to me from the very beginning, because the alternative scenario, in which my dad somehow, while driving, produces a book I hadn't noticed before and hands it over to me, seems highly implausible. My dad was never that smooth, and I doubt he had pockets that big.

trary to church doctrine nor (b) provoke your kids into asking any uncomfortable questions. My dad and I were far from the first Catholic father and son to face this conundrum, and for people like us, *Becoming a Man* was here to save the day.

As soon as he handed it to me,[12] I knew the worst had passed, as did he. True, the awkward car ride continued awhile longer; I think he asked me if I had any questions, and I managed to come up with some questions that fit in with the story we were both telling: that I had never once thought about anything sexual before this moment, but also that I had taken all the information he'd passed on and found it valuable. But we both knew it was just a formality. The talking part was over, my dad's job was done. After all, if there was one task I could be counted on to do as a child, it was read a book.

And believe me, I did. By the standards of Roman Catholic sex ed, it was pretty good.[13] But the main thing to me was that here, finally, was a book that would answer all the questions I wasn't allowed to ask. It might not give me the real answers, but it would give me the official answers. Which is exactly what I liked about it! I had stopped taking the Church's word as gospel[14] ever since my First Communion, when the bread and wine didn't taste like flesh and blood even a little bit. I knew that, when it came to the realities of sex, I was going to have to figure it all out on my own. But now I knew for sure what I was *supposed* to believe about sex. Never again would I express an opinion, or ask a question, that would make a grown-up uncomfortable, that would get me in trouble. Someone had given me the answer key.

Of course, my dad and I never spoke about that day again, and

---

12   Which, again, may or may not have been during this car ride.

13   It didn't waste any time pretending that teenage boys had the option to *not* masturbate, for example.

14   No pun intended?

I'll never know how he felt about it: whether he found it as awkward as I did, whether he felt proud or ashamed, whether he ever thought about it at all. But I have so much more sympathy for the dilemma he was in. I'm almost the same age as he was when we took that drive, and I can't imagine how I would handle it, given the constraints he was working under. And so, Dad, if you're out there: You did enough. You did your job. Thanks.

CHAPTER TWENTY-TWO

What Did You Win on *Jeopardy!*?

My *Jeopardy!* run entered the history books on January 26, 2022. It appears in the archives as a forty-game streak, running from November 17, 2021, to January 26, 2022, with winnings of $1,382,800, plus $2,000 for the second-place finish in game 41. But for myself, none of those statistics are quite right: when it started, when it ended, what I won.

I could say that my *Jeopardy!* run started in childhood, when my parents first began watching the "new guy," Alex Trebek, hosting a revival of a quiz show that had gone off the air a few months before I was born.

As for so many people, *Jeopardy!* has been in the background of my life for as long as I can remember—a calm, comforting routine: three contestants, sixty-one clues, three Daily Doubles. Potent Potables, Potpourri, "genre." Every weeknight, month after month, year after year. It was a place that valued the same things I'd been taught to value: curiosity, collegiality, just a hint of pedantry, and above all, a sense that knowledge was fun! I always believed that I would find myself there someday, and while I didn't know what would happen once I did, I knew I wouldn't regret finding out.

Or, of course, I could say that my run started when I first auditioned for the show, some fifteen years ago. As I recall, *Jeopardy!* had only recently begun administering their "entrance exam" online, in the form of a periodically offered timed test: fifty questions, fifteen seconds each. I was so excited. I crammed for a few days before, and (I'll admit) the first time I kept a few tabs open in my browser with some things I might be able to check within fifteen seconds (Oscar winners, that sort of thing). I did this every year; a few times I got to the final, in-person audition, and would live in hope for the next eighteen months, but I didn't get the call, and started again from the beginning.

But neither of those was truly the beginning of my *Jeopardy!* run. The childhood dream, that first audition, they were the beginning of something, yes, but that thing ended in 2017, when I realized I was trans. When I knew I could no longer deny my identity, and had no choice but to live openly as a trans woman, I closed the door on my *Jeopardy!* dreams.

Not because *Jeopardy!* wouldn't have had me on; I believe they'd already had trans contestants, but even if they hadn't, I knew the *Jeopardy!* ethos, and I knew that if a smart-enough trans person came along, they wouldn't hesitate to invite them on the show. But I believed then that I would never be comfortable enough with myself to go on national television, since at the time, even leaving my apartment felt almost existentially frightening.

But eventually I realized that, while I still feared that my appearance, my clothes, my hair, and most of all my voice would be disliked, both by me and by the viewing public, that fear no longer outweighed the happiness I knew I would feel at realizing my lifelong dream. I decided I was going to start trying out again, and that's when my actual *Jeopardy!* run began.

So, the next time the online test came along, I registered for it, although that brought its own dilemma. Should I use the same email address I'd always used, the one that would be associated in their records with my old name? On the one hand, I had a feeling that someone who had gotten close to being on the show a few times would be more likely to be selected. On the other hand, not only did I not want to be associated with my old name, doing so would also "out" me as trans. Even though I knew that being trans was no obstacle to appearing on *Jeopardy!*, and might even be a slight advantage,[1] that old fear still lived in me.

But I knew that, if I was actually to be invited on the show, it would become clear at some point that I was trans, and I was starting to feel a certain pride in my identity. So I decided to use the old email, "outing" myself as trans to them. Then I let that doubt go, and went back to reviewing lists of vice presidents, notable murder mystery authors, and Best Picture winners until the test came along.

When I got invited to the final round of auditions, I was excited, but didn't get my hopes up too much. I'd been there before. And for the most part, I felt ready to be seen: my hairline had been surgically filled in, I had quite a few outfits I actually liked wearing, and I'd gotten some confidence with wearing makeup,[2] but one problem remained: my voice.

Hormones can do a lot for trans women, but, if you've already gone through puberty, there are some things they can't fix, and one of the big ones is your voice. I had gotten some vocal feminization training, and could speak in a fairly passably feminine voice when I really concentrated on it. But I couldn't (and still can't) imagine using that voice all the time, every day for months or years until it became my natural speaking voice. To use that voice

---

1   They had no shortage of cis white dudes named things like "Tom" applying for the show.

2   Except for eyeliner, which even now intimidates me.

with my friends would feel fake, as if I wasn't really talking to them, but doing an impression for them, of a person they'd never met. And while I didn't mind being a little fake at work,[3] using it all day just seemed exhausting.

So I kept my original voice, but I still hated it. *Hated* it. I loved to sing as a child, but once my voice dropped in puberty I pretty much stopped singing entirely unless I was completely alone. When I went to mass with my parents, I would sing the hymns, but so quietly as to be almost inaudible.[4]

Years of cohosting a podcast, before and after transition, had helped me be slightly more tolerant of my speaking voice, but that only meant I could hear it with a kind of numbed indifference. And every time I was addressed as "sir" on the phone, a wave of self-loathing would come over me, sometimes for minutes, sometimes for days.[5] So, when I attended the "in-person" audition (this was 2020, so it was actually over Zoom), I used my "feminine" voice, and I resolved that if I got on the show itself, I would do the same.

And after years of effort, the call came: I was going to be on *Jeopardy!*

I was excited, of course, but still scared. The main fear was the one that had briefly caused me to give up, that fear of widespread hatred and scorn for being trans. I began practicing using my feminine voice for giving answers,[6] and planned out my first anecdote

---

3   Your workplace doesn't want you to be *too* real, after all. For example, you're supposed to pretend that you enjoy working there.

4   Although that was true of most of the congregation. And many of the ones that didn't sing under their breath probably should have.

5   This is still somewhat true. If you work customer service, and you see that you are talking to someone named "Amy," please don't call them "sir," regardless of their vocal register.

6   Sorry, "responses."

so I could practice the voice with that as well. As long as I used that voice, I told myself, maybe some people wouldn't even notice I was trans, and I might be spared the worst of the hate.

Due to various delays, my taping wound up getting pushed back by a year, and for nearly the entire time I practiced my voice off and on, and had every intention of using it on the show. But, just a few days before my taping was scheduled to take place, I realized that I had changed in such a way that using that feminine voice not only no longer felt like the safe move, it felt deeply wrong.

For one, I had realized that being trans in public carried responsibilities. I knew that many other trans people shared my vocal dysphoria, and that they had most likely never seen a trans woman with a voice like ours on-screen in any sort of sympathetic light.[7] I certainly hadn't, so to hide my voice began to feel like a betrayal of my community.

But I think the biggest reason I changed my decision was that, in the year between the call and the taping, I had started dating Genevieve. And Genevieve loved me, loved my body, loved my voice. And she made that so blatantly, undeniably clear that it put a big dent in that lingering self-hatred. If somebody as amazing and beautiful and perfect as Genevieve could love this voice, maybe it wasn't entirely bad? I still struggled with the decision, and I didn't finally choose until the actual day of the taping, but once I did, I knew I'd made the right choice.

And so there I was, in September 2021, with Genevieve driving me to the airport to fly to LA. The anxiety had been building for days. My hatred for myself had weakened significantly, yes, but

---

7 As opposed to, say, the 2009 episode of *The Jerry Springer Show* titled "Oops . . . I Had Sex with a Tranny!"

my fear of other people's opinions hadn't really changed. Would any of the *Jeopardy!* crew be weird about it? Would any of the other contestants misgender me? Would they object to my identity, or (almost as bad) ask me a bunch of intrusive questions about it? But by now I had mostly made my peace with that fear. The new, more acute fear had nothing to do with my identity: How would I feel afterward?

While I thought I had a good chance of doing okay at the game, it was a simple matter of statistics: Out of the ten people whose *Jeopardy!* journeys began that Tuesday, at least nine, and possibly all ten, would have their *Jeopardy!* run end that same day. So it was likely that, when I left LA, I'd be leaving *Jeopardy!* behind forever.

The dream that had started in half-formed thoughts as I played with He-Man action figures on the floor while my parents watched strangers answer questions[8] on TV, that had been part of the armor that carried me through adolescence (*Maybe I'm getting bullied now, but I'm a super genius who will win on* Jeopardy! *someday*); that had been a conversation starter at awkward social events with coworkers; that had been one of the few hopes I'd deemed precious enough to rescue from the collapsed heap of my old male identity . . . That dream, in all likelihood, was about to end.

Not only would it end, but its end would be seen by millions, an unknowable number of whom would commemorate the moment by echoing cruelties about my identity and appearance that I'd only barely been able to stop internalizing. After all that time, all I might get was thirty minutes of screen time, a few hundred bucks, and a smattering of insults on Twitter. I had worked hard to be okay with this, and I mostly was; after all, being on *Jeopardy!*, even for a single episode, is really fucking

---

8 "Respond" to "clues."

cool! But I still felt a dread that this trip would be an anticlimactic end to a lifetime of effort.

Of course, it wasn't! I defeated an impressive champion,[9] then won some more games, and then some more, and it became clear my run would last well beyond that Tuesday. But when did it end?

You could say that it ended on November 9, 2021, the day we taped the episode in which librarian Rhone Talsma ended my forty-game streak.[10] Yet at that point, as far as the world knew, my run hadn't even started.

You could say my run ended when my last episode aired on January 26, 2022, but that doesn't seem right, either. There was still the Tournament of Champions to come, of course, and given the success I'd had, it seemed probable that there would be even more *Jeopardy!* in my future after that. But even if you set that aside, my run as "Amy Schneider, *Jeopardy!* champion" is still ongoing, and probably always will be. For the rest of my life, I will forever be associated with the events of those few months, and it seems fair to say that my *Jeopardy!* run will never end.

I don't want to discount the monetary reward: a million dollars is a jaw-dropping amount of money to win by answering trivia questions, and for me and Genevieve (and Meep and Rue, the two most beautiful cats in the whole world), that alone has changed the course of our lives. But it's really only the beginning of what I've gained from this experience.

I've gotten to reconnect with all sorts of people from my past: old friends, teachers, castmates from long-ago productions of *Twelfth Night* and *The Lion in Winter*. I've been on the big screen at a Warriors game and heard applause from an entire arena. I've

---

9   With a bit of Final *Jeopardy!* luck.

10   With a bold Daily Double wager and a bit of Final *Jeopardy!* luck of his own.

gotten to appear on *Good Morning America*, talk to journalists I admire, have my picture in the *Washington Post* and the *New York Times*. My work was published on my favorite website.[11] I've gotten to spend time with the *Jeopardy!* crew, who were uniformly amazing people doing incredible work, and who I can't wait to meet again when the time comes. I've had all sorts of people and companies send me gifts, from chocolate chip cookies to designer clothing, and been recognized by the GLAAD Media Awards—all just for being myself on TV. And, of course, I got the opportunity to talk to you, through this book.

But before any of that, I got an even bigger prize. Until I appeared on television, I did not believe that I could ever be accepted for who I was. I had come to believe, not without some difficulty, that at least some people accepted me. But I still believed that those people were the exceptions, and that most people would see the things that I'd been raised to see in people like me: a freak, a pervert, mentally ill, ludicrous at best and evil at worst.

So, as the days counted down to my episodes airing, I braced myself for the rejection I was sure would come. And then . . . it didn't.

Sure, a few isolated voices popped up here and there to spew their hatred, but the overwhelming reaction was one of support and acceptance. It turns out, most people simply believe me when I say who I am! They accept me as I am, with my stubble, my voice, my thin hair. They don't think there's something wrong with me. And because of that, for maybe the first time in my life, I started to think there wasn't anything wrong with me, either.

I also know that my public acceptance isn't due to any special

---

11 Defector: the only sports-related website with a mixed-reality dugong.

qualities in myself, or at least, those qualities aren't the most important reason for it. The acceptance I've received is the fruit of long, violent struggles—some famous, some forgotten—in which trans people have risked their lives to secure their basic right to exist. Frances Thompson and Billy Tipton, Lili Elbe and Dora Richter, Sylvia Rivera and Felicia Elizondo, Laverne Cox and Gavin Grimm, and so many more, famous or forgotten, have devoted themselves to creating the conditions that exist today, where a trans *Jeopardy!* champion can be, for most people, uncritically accepted and celebrated as the person she is.

And the biggest prize of all that I received from my *Jeopardy!* run is the ability to say that I, too, have helped that cause. I haven't thrown rocks at the police, or fought for my rights in the Supreme Court; all I really did was chase a lifelong dream of appearing on *Jeopardy!,* and then answer a bunch of trivia questions.

But I knew that I would be taking on that burden of representation, and I will always and forever be proud to say that I've done my little part, helped ease the path for future generations of trans people to live free, open, and happy lives. That feeling is worth far more to me than any monetary prize, any media attention, any free gifts. It's something I will cherish forever.

But I'm still keeping the million dollars. As I once said to Ken Jennings, I like money!

Acknowledgments

First, I need to thank the people at *Jeopardy!* For the $1.3 million,[1] of course, and thus for giving me the opportunity to write this book in the first place. But more than that, I want to thank them for having been a quiet, unobtrusive force for LGBT acceptance for decades. I especially want to thank the people I worked with on game days: Corina, John, Laura, Parisa, Miguel, Sarah, Jimmy, and many others. They were the only people to be with me throughout one of the most intense experiences of my life, and they made it so much more fun than it could have been. Thanks to Ken Jennings; one of these days I'm going to retire from playing *Jeopardy!* just so Sony will allow us to be friends. Thanks to Mitch for the moments of calm. Thanks to Sam and Andrew, and the rest of the best ToC lineup in history. In particular, thanks to Mattea for representing, and for understanding.

Thanks to my mom for raising me to be curious and seek new experiences, and to see the value in the traditions that had been passed down to us, without judging the traditions of others.

---

1   And counting!

Thanks for knowing when to let go, when to let me find my own path through adulthood. Thanks for the stories you told, the gardens you took me to, the games of Poohsticks, the food I didn't appreciate enough, for my middle name. And thanks in advance for listening when I tell you which parts of this book I don't want you to read.

Thanks to my brother for reminding me what matters, for knowing that Cabbage Patch Bears aren't a thing, for being willing to forgive me, and for being one of the most interesting, kind, and fearlessly independent people I have ever met.

Thanks to Kelly, for being a great wife, a great ex-wife, and a great cohost. Thanks for getting me out of Ohio when I couldn't do it for myself. Thanks for ending our marriage when I couldn't do that for myself. I'm so glad you didn't get the part you wanted in *Cabaret*.

To the inner circle, Alex, Jess, and Candice: Thanks for the support, for the sunrises, for the gossip, for sticking with me. Thanks for listening to me complain about writing this book for the last year, and apologies for all the complaining when I write the next one.

Thanks to every trans person before me who refused to be silenced, who stood up to the contempt and the hate and the violence, who fought, and sometimes died, to bring all trans people out of the shadows, to force the world to accept our existence. I would not have had the courage that was displayed by so many bold, beautiful trans people. We have so many heroes whose names will never be remembered.

Thanks to all the people who helped me through my transition: friends, coworkers, bartenders, playmates, Twitter posters, YouTubers, strangers in bathrooms and locker rooms. But there are three people in particular without whom I don't know how I'd have managed. Jenn, thanks for being my trans mom despite, tragically, being cis. Thanks for complimenting my outfit that first time. Thanks for holding my hand throughout my transition, and listening to all my fears and doubts, even the dumb ones. Thanks for teaching me about boundaries, and calling me on my bullshit. Thanks for all the adventures, for all the people I'd never have met without you. Thanks to Melanie, who was a rock when everything else in my world was falling apart. Thanks for teaching me how to be single, and for showing me how to be a professional woman no matter what was going on in my life. Thanks for the shopping trips, the days at Mango and Emmy's, the happy hours, the hours on FaceTime when we were in quarantine. Thanks for getting me through. And thanks to Molly, who taught me how to be fabulous, how to work hard *and* play hard. Thanks for the compliments and fashion tips. Thanks for the tattoo.

It's somewhat obligatory at this point to thank the people who worked on this book with me, but my gratitude to them is both deep and sincere. I couldn't have asked for better people to guide me through this new world I found myself in. Thanks to Cait Hoyt, my agent, for reaching out, for believing in me before I did. Thanks for answering all my questions, for all the pep talks and patience. Thanks to Ben Loehnen, my editor, for staying confident in me even as pieces stayed "almost done" for months, as deadlines came and went without any perceptible results. Thanks for making me look like a writer, for taking out hundreds of adverbs and "thing"s, and thanks for leaving in the footnotes. Thanks to Ada Calhoun, who pushed me over the finish line, who saw the actual stories embedded in my stream of consciousness rambling, and whose explanation helped me finally understand what a "draft" is. Thanks to "Weird Al" Yankovic for allowing me to use a lyric from "Taco Grande" as my epigraph. When I was around fourteen years old, I decided that if I ever wrote a book, I would use that couplet as an epigraph, specifically because there would be no reason for it. I'm lucky to be able to fulfill one of my strangest childhood dreams. Thanks to Charlie Jane Anders, for classifying me as a "writer" before I thought I'd earned it. Thanks to Maggie Tokuda-Hall, for the insider info. Thanks to jasonrberwick on Spotify, for making "Best '90s Alternative Playlist Ever." Thanks to Brandi for the footnote. Thanks to Lisa Howard-Welch and Molly Shelestak for the photos.

I cannot find the words to express my love and gratitude for Genevieve. I don't really talk about her that much in this book,[2] but none of this could have happened without her love and support. The fact that such a smart, beautiful, cool, fun, silly, kind woman could love me is an ongoing miracle, and has given me a confidence in myself that I never thought possible. I may be better at *Jeopardy!* than she is, but she has far more wisdom, and I learn from her every single day. I love you, my silly little baby cat.

Thanks to Rue, for being so cute. Thanks to Meep, for being so fuzzy.

Thanks to you, for reading this book.[3] I hope you like it.

---

2   Because our marriage is none of y'all's business, is why.
3   Even the acknowledgments. I'm impressed!

About the Author[1]

Like Kim Deal, Paul Laurence Dunbar, and one of the two Wright Brothers, **Amy Schneider** was born in Dayton, Ohio. In 2021, she made her first appearance on the syndicated quiz show *Jeopardy!*, where she would go on to win 40 consecutive games, the second most in the show's history, trailing only Ken Jennings. She is the most successful woman ever to compete on *Jeopardy!*, as well as the only out trans person to compete in, and win, the show's prestigious Tournament of Champions. Since then, she has gone on to become a writer and LGBTQ advocate, has spoken at the White House, has been covered in a multitude of publications including *People, USA Today,* the *New York Times, Los Angeles Times,* and *Washington Post,* has appeared on *Good Morning America,* and has attended the White House Correspondents Dinner, where she saw Drew Barrymore in the women's bathroom but didn't introduce herself. Amy and her wife live in Oakland, California with their two cats, both of whom are extremely cute.

---

1   For further information about Amy Schneider, please refer to the entire rest of this book.